BREAKDOWNS IN COMPUTER SECURITY

Commentary and Analysis

Compiled by Michael E Rentell
Edited by Peter M Jenner

PA Consulting Group

Computer Weekly

Published by:

Computer Weekly Publications
Quadrant House, Sutton, Surrey, SM2 5AS
Publications Manager: John Riley
Deputy Publications Manager: Robin Frampton
Publications Executive: Katherine Canham

PA Consulting Group
Rochester House
33 Greycoat Street
London SW1P 2QF

ISBN 1 85384 024 6

A British Library Cataloguing in Publication Data catalogue record for this book is available from the British Library, London

Printed in England by Hobbs the Printers of Southampton

CONTENTS

Introduction	1
Computer Breakdown Incidents 1988	2
Computer Breakdown Incidents 1989	12
Computer Breakdown Incidents 1990	48
Generic Risk by Type	70
The Wider View	79
Index	91
Computer Weekly Publications	96

INTRODUCTION

Senior management personnel with responsibilities for data processing operations are required to protect the data their installations process against accidental and malicious destruction, interference or breach of confidentiality. The way in which such risks arise are many and varied and management also need to be kept up-to-date with any changes in these threats affecting their operations.

In order to address this need we have devised this book, which is a compendium of one hundred genuine incidents which have taken place in the past three or so years.

The details of each incident have been taken from freely available sources and should not be taken as representing any professional involvement on the part of PA Consulting Group or the authors.

As will be seen, incidents involving computers where victims have suffered serious, sometimes fatal, consequences happen surprisingly often. It is our belief that most of these losses could have been prevented had sensible computer security precautions been in place. We have, therefore, added to each incident a short comment indicating where improvements could have been made to alleviate or prevent the more damaging aspects of the problem.

We would be very happy to receive any details of additional incidents of this nature. The information would, of course, be treated with utmost confidentiality.

SEPTEMBER 1988

- The Federal Home Loan Bank in Seattle was spared serious problems when a major electrical fire left much of downtown Seattle without electricity for nearly a week. A newly installed disaster recovery plan was put into operation and the bank was back in operation within an hour. Other banks and 'Thrift' offices in the area were unable to proceed and needed armed guards to preserve their security.

  Even where they exist, disaster recovery plans are frequently neglected and are rarely exercised in full without notice. This small bank had taken the necessary precautions and the resulting competitive advantage is obvious. They will also have derived significant advantage in attracting/retaining business from competitors through the publicity of the event.

- The head of Soviet space exploration severely criticised mission controllers for a disaster which has disabled one of the two Russian space-probes on their way to the Martian moon Phobos. A controller mistakenly sent a message to the craft shutting down its rocket motor. This action in turn caused its solar panels to move out of alignment with the sun and all power was lost. There is now no way of retrieving the mission and the probe is effectively derelict.

  This case need not necessarily have involved a computer although it almost certainly does. The actual cost of this one mistake is obviously enormous. When working in expensive projects there must be control mechanisms, both in management and in machinery, which prevent single individuals either accidentally or deliberately issuing commands which can jeopardise the entire effort. In this case one would have expected the existence of system interlocks over such a vital system as spacecraft motor control. System interlocks ensure that every level of project management is in agreement before irrevocable vital procedures can be initiated.

- Two Soviet cosmonauts faced total oblivion when the computer on their spacecraft suddenly began processing an out-of-date program. This would have stopped the main rocket burning for the correct length of time during re-entry into the earth's atmosphere. They finally managed to get the re-programming done and landed a day later than planned - with only a further one day's supply of air left.

  This type of problem could have been avoided by any of several actions such as proper configuration control or more rigorous security (negative) testing. It demonstrates the requirement for the Quality Assurance organisation to be able to handle even those problems which are normally attributed to Murphy's Law: 'If anything can go wrong - it will, even if the probabilities of it happening have been assessed as negligible.'

- Over 150,000 records were destroyed on a computer operated by an insurance and brokerage firm in Fort Worth,

Texas after a member of the staff inserted a logic bomb which 'went off' a few days after he was sacked.

This has become a classic form of malicious attack amongst disenchanted personnel with access to the employer's computer. Poor program implementation controls offer the route by which the rogue program can be installed into the computer. This example shows how a simple act can cause major impact.

- An attempted Eurobond computer fraud which could have cost the London branch of Mitsubishi International Finance Corporation £20 million was prevented when operations staff at Shearson Lehman Hutton queried the unexpected receipt of funds. The criminals needed to gain access to EUCLID - the Eurobond information and clearing service. The bonds were transferred from Mitsubishi ownership to Shearson's by illegally accessing EUCLID from a personal computer and instructing funds to be collected in Switzerland. The perpetrators needed to be aware of Mitsubishi's passwords, but there was no implication that Shearson's staff were involved.

  This was an attempted line fraud using 'hacking' techniques. So long as EUCLID offers 'free delivery' facilities the security of its matched bargain procedures will not protect it from this type of attack.

OCTOBER 1988

- A whizz-kid type computer buff was discovered collecting lists of confidential passwords by bugging a network cabling junction box situated outside the building where the terminals were located.

  This example clearly illustrates the relative insecurity of simple user id/password systems for logical access control, especially when public lines are used.

- Four members of a stockbroking firm in France have been charged with embezzling 10 million francs using the company's computer. They were subsequently sentenced to 30 months jail.

  Most financial institutions operate a policy of dual authorisation for critical or high value functions. There is increasing evidence that such policies are not sufficient on their own. Without further details on this case, effective comment is difficult. Clearly there was collusion which always puts security controls at a disadvantage.

- The Prime Minister of Belgium, Mr Wilfried Martens, threatened court action because his electronic mail box had been raided consistently for three months by an unknown hacker. A newspaper published the story and printed a copy of the agenda for a government meeting as proof of the story. Mr Martens added that he would change his password - it had been 'W.M.'

Public electronic mail facilities must always be considered as insecure. Frequent changes of password and the use of complex strings of at least six characters are a must.

- Hill Samuel, the London merchant bank, narrowly escaped losing £60 million through an attempted wire fraud. They learned of the potential loss when their New York office informed them that their nostro account with them was in deficit. It was established that a series of phoney payment instructions had been drawn up the previous day and transmitted to countries with lax financial regulations.

This is not a true computer fraud. It is a wire fraud using the speed and efficiency of computer-based money transmission systems to beat internal control structures. It does, however, demonstrate the need to bring in appropriate controls and/or procedures which take account of the power of computer-based financial transaction systems.

NOVEMBER 1988

- Edward Austin Singh, a British hacker, wrote a host emulation program which he was able to insert into the telephone network and capture log-on passwords accessing Prime computer networks. Using these he was able to access computers on a world-wide scale. He was alleged to have penetrated over 200 sensitive military and commercial establishments. He was traced when he approached one of the world's largest computer firms and offered to explain how he had broken into their system. He wanted a fee of £3000 for this service.

  This is a skilled hacker using a well-known technique. Modern networks need to be fitted with sophisticated anti-hacking controls in order to detect this kind of attack at the outset. Once the host emulator program had been installed little more could be done and his downfall was his own fault.

- Recently discovered computer viruses have resulted in:

  - Electronic mail systems were severely disrupted at IBM by a Christmas Greetings message.
  - Personal computers at NASA lost data files.
  - Data files corrupted in computers owned by an electronic data systems corporation in Texas.

- The 'Pakistan' virus infected a newspaper office in the U.S.A.

- The largest network in Japan, serving nearly 50,000 subscribers, found a virus fraudulently charging network services to its users.

This month marked the beginning of a period of virus mania. Some were undoubtedly true viruses. Others appeared to be logic bombs or other malicious programs.

- Thousands of companies in the United States were temporarily overwhelmed by a computer virus as it spread through Internet a network used by scientific researchers. The perpetrator had exploited a ‘trap door’ in the Sendmail facility which had been left uncontrolled by the original programmer. The virus halted numerous research efforts, shut down networks and cost uncounted millions of dollars in clean-up time and effort. Later reports indicated that the perpetrator was Robert Tappan Morris Junior the son of the computer security director at the National Security Agency.

  A good example of the virus effect and of how an experiment can rapidly get out of hand. The existence of the trapdoor was the key and indicates that program testing routines operated by vendors could be improved. It also indicates how one can be terribly embarrassed by bright children.

- 30,000 computer microchips, worth half a million pounds, were stolen from a warehouse belonging to Macro-Marketing, Burnham, Bucks. The managing director of the company said that a world-wide shortage had created a black market in computer chips causing them to be worth their weight in gold.

  This incident appears to be purely a physical security problem but is indicative of how a rapidly changing market can alter the value of items previously considered unworthy of serious protection. This is often the case with today's 'hot commodity' information, where the controls in place are inadequately related to the value of the asset.

- Seven people were arrested after attempting to defraud the London branch of the Union Bank of Switzerland of £32 million on 29th June 1988. The money was transferred to the Nyon branch of Credit Suisse in Switzerland after the correct code and procedures authorising the transaction were entered into a computer terminal. The fraud was *accidentally* discovered during a manual check of transactions after a computer breakdown. The perpetrators were not employees of either bank.

  Full details of how this computer related fraud was perpetrated were never published. Either there was some password infidelity and the criminals exploited this or they used multiple tries and got lucky. In either case it is a worrying situation.

DECEMBER 1988

- A blaze at the offices of the Exxtor Group in Immingham destroyed its HP3000 computer and all its peripherals. Sensibly, the company had a disaster recovery contract, without it they would have been in severe difficulties. Even so, the replacement system offered 25% less power than their own machine and users were rationed to a limited number of terminals.

  Disaster relief plans are all very well but if they do not expand as the company's requirements grow then they are like a neglected insurance policy - when the crunch comes the compensation is not sufficient to avoid losses. The moral is that a recovery plan should be reassessed and upgraded regularly.

- In response to a disaster in Illinois when fire destroyed a telephone switching station, a new service providing cover against breakdowns in telephone systems has been made available in America. The disaster caused dozens of financial institutions to lose the use of the lines which carried vital computerised trading data.

  Spreading the communications load over a variety of carrier networks can reduce the impact of problems like this. However, in some countries, the national monopoly of communications adds to the difficulties.

- Three new computer viruses infected companies across the UK. One was primed to go off at an unspecified date in the future and one which was the first found to be in an encrypted form and which appeared to be restricted to Scotland.

  The instances of virus attacks will undoubtedly continue to increase and the perpetrators will become more sophisticated. Our defences will have to improve accordingly. This is destined to grow as a problem until we have effective laws against it.

- The Futures Exchange in New Zealand had arranged for Reuters to operate the back-up facilities for its computer trading operations and then did not use the facility on a day when power problems halted proceedings. As a result, futures traders were done out of around $2 billion of business and half the purpose of the futures market - hedging and physical market - was defeated.

  This is a management problem. The explanations given for why the established back-up system was not used are somewhat convoluted. Essentially the management never envisaged losses to be so great and therefore the entire incident can be classed as the result of insufficiently detailed planning.

JANUARY 1989

- Students attending the IBM training centre at La Hulpa in Belgium returned to their various employers with diskettes infected with a computer virus contracted at the IBM site.

  This case shows how even the largest and most powerful organisations can fall victim to computer virus attack. Even if actual losses are minimal the effect on the company's credibility can be extensive. 'Et tu Big Blue!' is a cry which probably rings sweetly in the ears of IBM's competitors.

- A man has been arrested in Los Angeles for using unauthorised telephone access codes in order to hack into computers at the University of Southern California.

  A simple case of hacking. In many countries this activity remains legal. Only in some states of the USA is it an arrestable offence. However, times change and the British Law Commission has recently presented a White Paper to the government advocating a change in the law to make hacking an offence - although a distinction is made between hacking just to gain access and hacking with prior knowledge and intent to cause harm.

- The Livermore National Laboratory in California, America's premier computer research centre, has suffered at the hands of a hacker to such an extent that the management has asked the intruder to come forward. Defences against the attack have so far cost over $100,000 and all in vain.

The centre security manager stated that the hacker acquired 'super user' status with facilities to delete any information he likes. However, it is believed that to date he has not actually deleted any files. It is also believed that he has not breached classified areas of the computer.

Adequate defences require that the necessary access controls were erected in the first place. This case shows how difficult it is to combat a determined and skilful hacker once he/she has gained access to the system.

FEBRUARY 1989

- Virus writing kits are currently being manufactured for hackers by members of the Chaos Computer Club in West Germany.

  This is the most notorious federation of hackers. At the time this occurred hacking activity was not against the law anywhere in Europe.

- More PC viruses now positively identified include:

  - '1701' & '1704'. Sometimes trigger hailstorms on the screen.
  - 'Brain'. Some processing is slowed down but otherwise no other effects yet seen.
  - '1813'. Some processing is slowed down and, more damagingly, deletes every program run on Friday the 13th. Sometimes opens a small black window in the screen's bottom left hand corner.
  - 'Stoned'. This displays 'Your computer is now stoned' and corrupts some hard disks involved.
  - '648'. Some computer programs will not run.
  - 'Italian'. Creates a bouncing dot on the screen and other side-effects.

As with any interesting form of vandalism, there is a bandwagon effect. These viruses are a result of publicity of earlier examples.

- A loan servicing company hired by the California Student Loan Finance corporation failed to meet the federal government's servicing requirements due mainly to a massive computer failure. As a result, the education department is refusing to guarantee much of the $1 billion portfolio amassed by the corporation. Now some of the world's largest banks are threatening to sue the corporation's trustee bank, claiming the company did not adequately monitor the servicing of the loans.

  This is a management problem and as such did not happen overnight. There should have been properly engineered contingency plans and regular capacity management enquiries which would have indicated a build-up to overload conditions.

- British Rail experienced a computer virus attack on Friday 13th but services were not seriously affected. The virus was isolated and dealt with shortly after its discovery and was revealed as the '1813' virus.

  The predictions of chaos to be caused by this virus on this day were wildly exaggerated. British Rail should be congratulated for dealing swiftly and effectively with the problem, although proper management controls should have ensured that they were not infected in the first place.

- In a report on computing in company accountancy the three most likely causes of disaster were listed as:

> - Failure to train oneself and staff in the necessary disciplines when moving from non-critical applications such as word processing to those at the heart of the business such as invoicing and keeping track of money.
>
> - Computerising accounts when neither the boss nor those who will be running the system understand accountancy sufficiently to make effective use of any package.
>
> - Selecting the wrong accounting package.

The last is believed to be the most common and fastest route to disaster.

These are surprising conclusions but valid nevertheless. All three points are management problems and are based upon insufficient knowledge at senior level. In our experience it is often the case that management awareness of computer related problems falls far short of what it should be.

- A computer virus 'vaccine' program which was supposed to delete the '1813' virus from infected computers was marketed by a well-known PC guru. Very shortly afterwards it began to be circulated carrying an additional part which actually scrambled any computer disk it came into contact with.

This is a dire warning against using copies of useful programs or even 'pirated' software. Companies should never need to be in such a position. If there is any doubt whatsoever, such programs should be confined to stand-alone machines until they have been proved to be virus free. The currently fashionable terms for this procedure is 'computer hygiene' or 'software quarantine'.

- A 17 year old bank cashier from London used his branch computer to defraud a major retail clearing bank of nearly £1 million. He and his friend, into whose account at another bank he transferred the money, were each convicted of conspiracy to defraud. The judge in the case refused to make a complete restoration order in the bank's favour because the system was so insecure that 'general chat' in the branch seemed to indicate how easy it would be to carry out a fraud.

This is an unfortunate illustration of the state of affairs in an organization which generally takes computer security seriously. Once again this shows that the greatest threat comes from 'insiders'. This is a management problem and the judge's actions 'pour encourager les autres' should help to drive the point home at a senior level.

MARCH 1989

- West German computer hackers repeatedly tricked the Soviet KGB into paying up to DM300,000 for easily accessible and strategically insignificant computer-based information. The hackers exploited the Russians' ignorance and walked away with a tidy profit. Targets were the public data banks of the Pentagon, Thomson SA and the Max Planck Institute for Nuclear Physics.

This is a rather nice illustration of an incorrect assessment of the 'value' of information. It also indicates how easily publicly accessible networks can be attacked by unauthorised persons.

APRIL 1989

- Experts investigating the Exxon Valdez oil disaster off Alaska are trying to find out exactly when the computerised auto-pilot system was switched off. One theory which has been put forward is that the auto-pilot had not been shut down and the crew's first attempts to take avoiding action were overruled by the system.

  When computers are placed in charge of a real-time control system the program specification needs to be extremely detailed and must take into consideration every possible scenario. Another example of computer programming acting in a detrimental manner is the automated selling at trigger points of stocks and shares which can lead to major market crashes.

- The Companies Registration Office in Cardiff has decided to double the resources reserved for computer back-up in case of disaster. Within the last seven months they have suffered 35 computing breakdowns. Of these, four had been due to mainframe crashes, two were due to loss of power and there were operating systems problems in two other instances. The remaining 31 instances were communications problems.

  Most of the problems can be attributed to bad communications links stemming from the unexpected increase in public use of the office's available on-line search facilities. This is basically a capacity planning problem although poor communications facilities are also involved. It serves to illustrate how the value of a major investment can easily be lost through not considering all potential use scenarios.

- An earthquake in Mexico City affected the stock and money market computer systems. Data and telephone connections remained affected for many hours.

  In an unstable environment it may be considered unwise to commit large resources to sophisticated computer systems when they could be knocked out at any moment. The chances of establishing reliable back-up systems in the immediate area must also be considered as unlikely. Computing science at the moment really has no answer to earthquakes (except perhaps remote computing and satellite links).

- The launch of the space shuttle Atlantis was postponed for 24 hours because of computer problems. Initially a computer fault halted the count-down for five minutes but eventually the launch was cancelled because the fault which caused the failure could not be identified.

  Swift and accurate identification of faults in complex computer systems needs to be engineered into the system at the initial specification stage.

MAY 1989

- Millions of people who opted out of the State Earnings Related Pension Scheme and took personal pension policies just before the April 1989 deadline are being sent policies and cooling-off notices that show totally incorrect benefit entitlements. It was reported that insurance company computer systems could not cope with the mass of detail by the government imposed deadline so programmers were guessing at benefit levels which will need to be changed later.

  This is basically a result of programming inflexibility into the systems. Sometimes it is necessary to use a little imagination and do the guess-work before the event rather than afterwards.

- A firm of typesetters in Buckinghamshire staggered from crisis to crisis with its computer equipment until February 1989 when an investigation revealed that the system was harbouring the Lazy Harry virus. All the failures and distortions of screens had cost the company nearly £20,000. The virus had been imported by the son of a senior partner who had used the computer to run a games diskette borrowed from his older brother.

  This incident shows how important it is to ensure that only authorised software is run on computers processing vital information. Using critical systems to run games programs can, and in this case did, lead to disaster. Games programs are one of the main propagating media for computer viruses.

- Inadequacies in specialist software which prepares export documentation is reported to be costing the UK's top exporting firms significant losses through delays and export rejections. The software is alleged to offer an inadequate range of form layouts which can cause the user to send incorrect documentation with exported goods. The result of this is often rejection by customs authorities in the distant land.

  The adage which should be repeated whenever contemplating the purchase of off-the-shelf software is 'Try before you buy'. Programs which provide services critical to the functioning of a company must be tested thoroughly before they are implemented to ensure that all the requirements are available.

- A burst water main flooded the main computer room of Extel Financial, the information service to the City of London. It was forced to suspend its on-line information feed and restrict its stock price service to morning newspapers. The burst pipe was detected by automatic monitors which closed down the computer systems immediately. Water was found to be an inch deep in the cavity under the computer room floor. The water pipe was part of the environmental control system maintaining the computer room at a steady temperature.

  The detectors worked, the automatic shut-down worked, so the plans seem to have been well-laid. It is perhaps more important to ask why there was no automatic switch over to a back-up system and whether the possibility of such an incident had been planned for.

- A total systems failure at a major power tools automated national distribution centre resulted in the complete destruction of its entire database *and the back-up copies thereof*. There was no easy method of accessing the stock for distribution and eventually the staff were reduced to wearing mountaineering equipment in order to scale the roof-high stock bins to do stock-taking. The costs were not just the losses involved in the reduced inefficiency, but also to the company's established goodwill. The root cause of the problem was never fully identified but it seems that an intermittent cable fault initiated a chain reaction which resulted in a partial hardware failure.

  This was a genuine disaster and the company admits that the costs are incalculable. Clearly the loss of back-up data should not have occurred and the lack of any contingency plan appears also to have been a serious problem. Copies of critical data should always be kept off-site.

JUNE 1989

- Six City of London firms are reported to have signed agreements with computer hackers offering them amnesty if they return some of the money stolen through hacking. In some cases, losses exceed £1 million but only a tenth was returned. The firms are apparently willing to accommodate hackers because they fear their business credibility could be affected if the facts became common knowledge.

  This story comes from the Computer Industry Research Unit. If it is true the situation is a sad reflection of the state of computer-based security procedures. Detailed procedures are available now which will ensure that hackers cannot enter financial systems and 'steal' funds.

- Two technicians, once employed by a computer marketing and installation company, have been arrested in Barcelona and charged with infecting municipal computers all over Spain with a virus. They then offered to sell town councils the relevant 'vaccine' program.

  A typical tale of computer-based extortion. This, once again, reveals that unprotected networks can succumb to the 'virus' threat very easily. Proper security controls, adequately managed would go a long way to ensuring that such crimes could not happen.

- A blaze at the Clerical Medical insurance company headquarters in Bristol caused severe smoke damage to the company's mainframe computer and associated minicomputers. Fortunately the company had all data properly backed-up and could transfer its main operations to a similarly equipped programming facility on the same site and to a disaster recovery site at the computer vendor's location. The company had practised for such a disaster only six months previously and a spokesperson said the experience had clearly paid off.

This company deserves all the good fortune they experienced. They had planned for such a disaster and, equally importantly, had exercised their staff accordingly. If all computer users did likewise the losses from such fires etc would be much reduced.

- An enquiry was told that police tried desperately for several minutes to persuade the fire brigade to send vital cutting equipment to the Hillsborough Stadium in Sheffield when the football terrace became disastrously overcrowded and 95 people died. The fire service operator said that the computer would not recognise the Hillsborough football ground as a place - 'it needs a specific address and district in order to determine what specific pumps are required to attend.'

Hindsight, they say, is an exact science. However, we should perhaps ponder upon how seemingly simple programming decisions can have very serious consequences when implemented.

- The High Court in London has frozen £250,000 held in bank accounts in England which the Bank of Crete states was allegedly transferred from it as part of a massive fraud perpetrated by the bank's former chairman and his brother. Documents had been shredded and computer records of deposit accounts were alleged to have been rewritten putting them in the chairman's name. At the time, deposit accounts were immune from scrutiny under Greek law.

  This is a classic computer-assisted banking fraud. Poor controls of computer resources plus apparent collusion at a senior level are at the root of this alleged crime. Here again we may find evidence that collusion is a problem to be guarded against.

- The computer centre operated by one of Britain's biggest building societies experienced a breakdown when a back-up generator failed a pre-test check. The generator at a back-up site failed to start in a disaster recovery procedures test. The test showed that electricity supply had been low down on the list of potential risks to the centre.

  The society's planning staff should be congratulated for carrying out the pre-test check and establishing the problem. So many contingency plans are never properly exercised and when there is a genuine emergency all sorts of problems arise.

- Twenty five per cent of all computer professionals questioned in an anonymous survey carried out by the

Computer Weekly magazine have admitted to hacking. Of those admitting gaining unauthorised access to their computers, 87% reported that they were never found out.

There are various ways of defining 'hacking'. We would suggest that the current interpretation of the word - 'forced logical entry over a network without knowing the password' - is unlikely to be the proper term in these cases. The new legal definition of 'gaining unauthorised access' is perhaps more relevant. However, the claim that attempts to enter the system were not detected is much more likely to reflect the true situation at the majority of computer centres.

JULY 1989

- In an effort to prove that its disaster policy is viable, a large company arranged for a special team of professionals to simulate a disaster. The team arrived with signs which were attached to equipment, people and rooms signifying that they were out of action. The rest of the staff were then required to write out what they would do to adjust for the losses. One program simulated an earthquake and the head of data processing was required to stay in the men's toilet with a simulated head injury until found. No one missed him for four hours.

  Disaster planning is a necessary part of operating a key computer installation. Whether one should go in for quite such a detailed exercise is debatable, particularly if it reveals personally unwelcome facts like this.

- Delegates at a computer disaster symposium were informed that in the event of a disaster in a computing operation senior executives could find themselves held personally liable for the loss of data because the law requires that management maintains and safeguards all corporate records on behalf of shareholders.

  This situation, once it becomes common knowledge in board rooms, will help companies generate the will to provide reliable computer disaster and data protection plans.

- Spanish tax authorities collected the computer records of a quarter of a million savers with a bank who had invested in insurance policies which were marketed as tax free. The bolt-holes for proceeds of the 'black economy' had been thought secure under banking secrecy regulations, but revenue officers succeeded in impounding the relevant computer disks under tax evasion regulations.

  The data density of modern magnetic storage media ensures taxation officers catch many offenders in a single trawl. It also serves to remind management just how much data could be lost if the medium is damaged.

- A travel agency has gone out of business because hackers booked up hundreds of airline seats which were then never paid for.

  This could either be a deliberate attack by a competitor or even possibly organised crime operating a computer extortion racket. Whatever the case, the attackers clearly knew the system well enough to recognise that whoever perpetrated it would not be identified. When planning such a system the openness which is its main attractive feature has to be balanced against the known threat of malicious damage.

- Police in France received evidence that hackers gained access to medical databases containing information on AIDS victims and used it for blackmail.

Once again this is a case of an open system being exploited for criminal gain. This case proves the need for governments to insist on the installation of efficient network security controls over public systems.

- An estimated 25,000 120 megabyte disk drives for use in PS/2 Model 70 desktop computers were recalled by IBM after the company concluded that they would fail due to leakage of lubricants.

  Insufficient component testing during the product design phase is probably at the root of this problem. This incident reinforces the need to take regular back-ups of critical data.

- Recent instances of computer related problems in the medical field include:

  - Hackers got into a computerised intensive care system over a network and changed some critical threshold settings which were set to trigger an alarm when the patients' conditions deteriorated. Two patients died.
  - Other hackers gained access to computerised hospital medical records and altered them.
  - A cancer patient was blackmailed when hackers gained access to computerised medical records and discovered her condition.

- The computerised drugs database at a major British drug manufacturer was entered by hackers and tampered with.

- An attempted murder by an outsider trying to administer a fatal drug overdose by using a doctor's personal computer password occurred at a hospital in Albuquerque in New Mexico. The victim was saved by a quick-witted senior pharmacist who recognised the code and knew that the authorised holder was not in the hospital at the time.

Now that computers are used in real-time medical control systems the opportunity for remotely-controlled murder and other crimes is now apparent. Controls must be introduced at the system design stage to ensure that such activities are not possible.

- Storms in England caused damage to computer cabling systems from connections on the outsides of buildings. Where cables were not properly earthed the lightning travelled through the buildings and damaged terminals and computers.

Proper attention to wiring safety techniques is essential when installing computer systems. When lightning strikes there is a considerable electromagnetic field induced in nearby wiring and this is quite capable of destroying modern CMOS-based chips in nearby electronic equipment even if it is not connected to it.

- American lawyers, retained by relatives of passengers killed when the American warship USS Vincennes shot down an Iranian civil airliner over the Gulf, are claiming that the Aegis Mark 7 weapon system computer was incapable of distinguishing between a civil airliner and an fighter plane.

  Whether or not this claim is accepted remains to be seen. However, the very fact that lawyers are prepared to make the claim serves as a reminder that specification and program testing are of the utmost importance. Nowadays it seems that even the most unlikely eventuality needs to be included in program specifications in order to ensure ultimate reliability.

- A computer operator has admitted accidentally deleting tapes containing thousands of computer processed documents concerning the Exxon Valdez oil tanker disaster off Alaska. The documents were subject to a court order requiring their preservation. However, the tapes were not properly labelled and the operator believed he was simply performing routine 'house-keeping' duties. He was sacked.

  This is a management problem. The proper labelling of vital tape spools is an obvious requirement, particularly when the data owner may suffer financial penalty if they are destroyed. One cannot help sympathising with the individual concerned.

- A Swedish bank has admitted that it recently suffered an attempt to steal £57 million via an international computer network.

  As usual in such cases, the details are very sketchy. Almost certainly this was an inside job with no breach of the international electronic payment system security controls. No matter how tight such controls are, if the clients do not apply the recommended procedures assiduously they will sooner or later end up the loser.

- A development program to reduce the risks of failure in the unstable computer systems at the Australian Stock Exchange will take a year and cost $30million. There were three failures in one week which forced the automatic data feed system down for two afternoons and resulted in the automatic trading system being unavailable for two hours.

  Trying to force a system to operate satisfactorily beyond its capacity in order to save money is really a false economy. This report indicates how important it is to keep ahead of capacity planning requirements and to budget for systematic expansion when indicated.

- A computer virus nicknamed 'Datacrime' was discovered which was reputed to crash infected machines on Friday 13th October 1989.

  A similar warning was also circulated at the end of 1988 concerning Friday 13th January 1989. The extent of the infection then was very limited within the United Kingdom.

This warning was timely enough to allow for the programming and marketing of suitable anti-virus software.

- Faults in the software have delayed a multimillion pound computer project sponsored by the Meteorological Office to improve weather forecasting and research into climatic change.

  Complex problems require complex solutions and underestimating development costs is a common problem. Buying the right hardware for the job helps at the outset.

- Fraud involving computer-based currency rates in a large investment bank in France is reported to have caused losses of around FF3.3 billion and the bankruptcy of the bank.

  We will probably never know the details of this report but the lack of central control is likely to have been a major contributory factor.

AUGUST 1989

- Five former employees of British Airways are pursuing claims for compensation for injuries allegedly received as a result of prolonged use of computer VDU terminals.

  Such problems rarely feature in the plans of manufacturers or users. The matter has now been investigated by the Health and Safety Executive.

- Three West German hackers have been charged with spying for the USSR after allegations of breaking into secret Western computer systems.

  This is a result of the item reported in March 1989 (see above). The skills of hackers, particularly in West Germany, are becoming so extended that their activities now appear to constitute a threat to NATO security.

- A small British engineering company claimed to have lost nearly £1 million due to defective software supplied by an American software development house. It was claimed that the software failed to meet specification.

  Adequate program testing is vital. However, even thorough testing will not identify problems if the original specification was not sufficiently detailed.

- The US General Accounting office cited the lack of any central reporting agency across the US research network

Internet as being partly responsible for the computer virus which affected the network in November 1988 (see above). When it was discovered no one knew whom to report it to and how to stop it.

Once again a management problem surfaces as the root cause of a disaster. When this loosely controlled network was initially commissioned someone should have been made responsible for its operation and all users should have been made aware of the fact.

- Police have admitted that recently introduced copyright legislation is ineffective in dealing with software theft. They have had to drop a 'conspiracy to steal' case against an individual involved in the illicit copying of computer object code. If the printed source code had been stolen there would have been no problem. The case is doubly interesting because it involved an apparently innocent party - a supervisor at a user's location who was instructed over the telephone to carry out certain actions on her computer by the engineer responsible. These instructions, which she did not understand, apparently opened a 'trap-door' in the program which allowed it to be copied. The software owners stood to lose a large sum of money invested in developing and marketing it. In view of the failure of the criminal process they are considering a civil action.

  It was anticipated that the new copyright legislation would be sufficient in this case but clearly it is not. Law makers are still apparently unable to come to terms with modern technology and devise laws which adequately cover the devious and obviously undesirable activities of the astute and technically aware criminal.

- A well-known American hacker has reported that, under duress, he was required to instruct other jail-birds in the techniques of hacking whilst he was serving a sentence for the offence. For several months he instructed drug-dealers and sundry other felons in the principles of telephone operated frauds.

  As technology becomes more complex and its criminal exploitation becomes more sophisticated, this sort of claim is likely to be heard more often. In this case the individual was using his alleged experience as an argument against framing laws permitting the jailing of malicious hackers. Such a specious philosophy could also be cited by cases of criminal locksmiths and burglar alarm manufacturers.

- A graphics-based computer game which is normally innocuous has appeared with 'conversation bubbles' peppered with French obscenities. The UK distributors of the game were required to withdraw it from circulation by retailers in England. The manufacturers are at a loss to explain how it happened.

  If the copies which contain the obscenities are genuine manufacturers' originals then there is clearly a testing problem at their end. One wonders, however, if the offensive diskettes in circulation are not 'pirate' copies which have been tampered with by someone with knowledge of manipulating hidden data files.

- The new computer installed by the Garda Siochana (police) in Eire was knocked out when lightning struck the uninterruptible power supply building. The computer serves all police stations in the Irish Republic and deals with all 999 calls. It was out of action for several days whilst repairs were made. Delays were caused by unfamiliarity with new equipment.

  When developing contingency plans for the sudden loss of computer power the use of an uninterruptible power supply is of great value. The mains electrical power supply is likely to be the first affected by lightning strike and so it is important to ensure that the standby supply system is able to withstand such an event.

SEPTEMBER 1989

- Over 40,000 car drivers in Paris received computer generated notifications which indicated that their motoring misdemeanours had mysteriously been interpreted as murder, prostitution or extortion. They were subsequently notified that a computer 'error' was responsible and received an apology.

  This is a case which causes a smile amongst those not involved. However, it is a serious matter and one which the excuse 'computer error' is difficult to believe. Could it be the computer centre suffered a malicious attack from a disgruntled employee? This is precisely the sort of occurrence which indicates such an attack.

- Docklands Light Railway in London was halted when the electricity supply failed despite the existence of uninterruptible power supplies and a back-up computer. Passengers were stranded for over an hour as the computer-controlled systems were investigated. It appeared that the standby power supply was unable to maintain supplies for more than a short period. However, this should ensure that the trains continue to run but the mains supply was so erratic that both systems failed. The failures in supply were claimed to be a 'million to one shot'.

  Although the railway authorities had planned for losses of power, their system was unable to cope with the power going off, on, off in quick succession. This underlines the necessity of taking a wide view of possible disaster scenarios when planning back-up systems.

- A jury at the Old Bailey was told that a computer operator aged 20 had approached a competitor of the company for which he worked and offered to sell his employer's computer-based secrets. He was jailed for 18 months. Had he been successful it would have been disastrous for his employer.

  Once again a serious threat is posed by an 'insider'. There are serious risks involved in allowing immature staff uncontrolled access to vital computer resources and data.

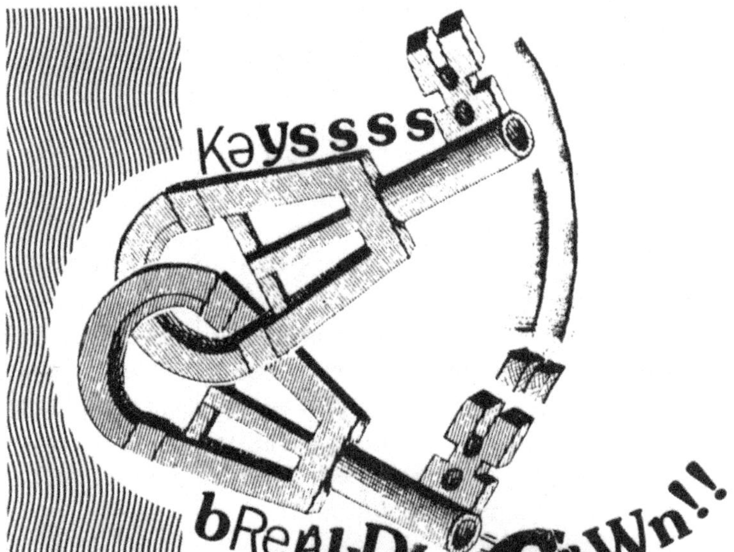

OCTOBER 1989

- The expected rash of computer viruses on Friday 13th October 1989 did not materialise in the United Kingdom although there were several outbreaks around the world. The dangerous virus *DATACRIME* was reported in Switzerland and in Holland where up to 10% of PCs were thought to be infected. This required a massive effort by police in making 'vaccine' software available to anyone who called and reduced the problem to manageable proportions. In Portugal an unidentified virus attacked an industrial company and a bank.

  All Friday 13th dates are now attracting virus planters. Armed with such knowledge it is possible to prevent most cases of infection by (1) ensuring good computer 'hygiene' (2) not allowing pirated or second-hand software into PCs and (3) investing in suitable 'vaccine' software.

- A malicious program which worms its way from one computer to another over networks has been found in the NASA Span network and in Hepnet used by high energy physicists in America. It announces that it comes from 'Worms Against Nuclear Killers' and leaves the message 'Your system has been officially wanked'. As a 'WORM' program this effect is generally unremarkable. What is so special about it is that this program is not based on PC networks but only works on VAX computers manufactured by The Digital Equipment Corporation and there are many such machines on these networks.

Its run-time file name is W.COM which indicates that it is not a true program but a DCL, a file of control commands used for machine scheduling. After displaying its disconcerting message, it alters the system messages to indicate that all disk files have been deleted - but they haven't. It then modifies the privileges of the system supervisor so that no electronic mail messages, which could contain warnings about it, can be generated. It then searches for user records without passwords and for passwords which are the same as the corresponding user-ids and sends the information back to the mysterious originator thus effectively creating 'trapdoors' for later exploitation.

The program's final act is to search the dial-up lists of all other computers on the network, telephones them until it finds one on-line and then transfers a copy of itself to it where the process is repeated. This final activity effectively makes the program a 'virus' and therefore represents the first genuine mainframe virus generally notified and publicised.

The arrival of a mainframe computer virus has been predicted by PA for sometime. Now, finally, one has been let loose on a widely used telephone-based network where it can do a lot of damage. The fact that it is a series of DCL commands indicates that it has probably been created by an astute systems programmer. The deletion of this program from affected systems is obviously a requirement, but also DEC users must now ensure that they have got properly constructed account passwords and that their passwords are not the same as their user-ids (account names).

- In what appears to be the first reported case of computer crime in the People's Republic of China, a bank in Chengdu has reported that computer criminals recently stole the equivalent of $235,000.

  Details of this incident are sketchy but it clearly indicates the world-wide nature of such problems. When the target is a large sum of money there is no suppressing the ingenuity of people.

- Building societies and banks are finding it difficult to maintain their levels of service to mortgage borrowers because over a year after multiple tax relief on mortgages was abolished their computers are still not able to protect borrowers' interests. Where two individuals are in process of buying a property jointly and one then moves out the expected tax relief has not been forthcoming. The reason given is that the societies have been so busy over the past year bringing out new products they they have not kept pace with legislative requirements.

  Market driven computing effort can often cause problems. Whilst legitimate program maintenance is subordinated to market forces such problems will always occur.

- Only one case of Friday 13th virus infection was reported to have struck on the fateful day, at the Royal National Institute for the Blind's employment technology unit. It appears that the virus was of the '1813' variety and was not the dreaded 'Datacrime' which, so far, has not appeared in the U.K.

This day was trumpeted around the media as the day when the nation's computers would grind to a halt. It is a pity that the only organisation to report an infection was a registered charity. The warnings of such infections are now clearly having an impact and organisations are taking some action to limit infection and consequential damage.

- A London bank operated its APACS computer-based payment system twice in one day and succeeded in transmitting £2 billion more than it should have done to the payment agency. Luckily the error was spotted and all the funds were successfully recovered.

  This is a procedural error which should have been made impossible by proper programming and operating procedures. It is interesting to note that the bank was not identified although a statement was published which indicated that the proper programming action has now been implemented and that such a 'double run' cannot happen again.

- A computer-based fire control system has ruined a restaurant owner's attempt to build a record-breaking pyramid of hot samosas. The attempt, which took place in Durban, was spoiled when the computer-based sensors detected the excess heat and turned the water sprinklers on.

Such a situation was clearly undreamed of when the computer programmer was given the specification for the heat monitoring program. It seems unlikely that 'if heat detection equates to 3 tons of samosas then do not turn on sprinklers' could have been foreseen.

- A piece of American software designed to increase the bill of cellular telephone users is being touted around retailers of such equipment in Britain. It permits operators to identify the most commonly dialled numbers users access and to increase the recorded number of times these are used. It relies on the fact that most users do not record exactly how often these frequently used numbers are dialled over any billing period. British retailers are extremely disturbed by the appearance of the software which does nothing for their image.

As electronics based systems become more complicated they increasingly rely on programmable resources for system management. Once a computer program is involved with money then the opportunities for fraudulent programming arise. Publicity is one way of ensuring that people are aware of such criminal activity and are on their guard against it.

- Collated information concerning the San Francisco earthquake reveals several instances of computer-related losses:

> - There was a power cut lasting over 30 hours which brought newspaper production to a halt. Two of the most well-known newspapers had prudently invested in back-up generators but as loads had increased since their installation they were inadequate for the task and could not support the mainframe computers which were vital to timely newspaper production.
>
> - The airport was closed for 13 hours because the control tower computer had fallen out of a window as a result of the shaking.
>
> - There were serious data losses at San Francisco City Hall when the power to the computer data disk drives was lost.

Planning for major earthquakes may be considered somewhat esoteric in the more stable parts of the world but on the California coast it is a must. The ability to close down computer operations in a controlled manner needs to be considered in all computer-supported enterprises in such areas and auto-start back-up generators with uninterruptible switch-over should be considered as vital.

DECEMBER 1989

- An 'AIDS INFORMATION' package in the form of a computer readable diskette was mailed to over 10,000 business addresses in Britain. Instructions with the diskette required it to be placed into any IBM compatible PC and run in the normal way. Once this happened a 'Trojan Horse' program copied itself on to the computer's hard disk and displayed a message stating that the contents of the disk would be destroyed unless a licence fee of between $189 to $378 was paid to a company in Panama. The odd thing about the case is that the accompanying documentation clearly contained warnings about the threat although in very small print.

Judging from the number of individuals who complained about the damage this program threatened, it is clear that they either did not read the warning or did not heed it. Although police are viewing this case as extortion, this method of ensuring the payment of software licence fees has been seriously mooted in the past. Perhaps the perpetrators considered that including a warning in the accompanying documentation exonerated them from all the consequences.

JANUARY 1990

- A massive fraud using the computer-based money transfer systems used by a London-based retail bank led eventually to bank clerk Stephen Hedges, 23, being jailed for three years. The total fraud exceeded £1.5 million and was not noticed by the bank for nearly a year. Hedges simply registered a false company, transferred the money in five transactions and destroyed most of the evidence. He then left the bank, transferred the cash from his company to a Swiss bank account, and set about a life of Reilly around the world's most glamourous resorts. The auditors finally learned of his actions and he was caught whilst trying to run away with the money to Bangkok.

  Many people would call this a Computer Crime. This is almost certainly not the case, but the speed and efficiency of computer-based systems were exploited in order to get the money which had been diverted by a manual fraud from one place to another. It may be that Hedges never actually touched a computer in perpetrating this fraud, but he was certainly aware of its potential in his scheme of things.

- The giant American communications corporation AT&T suffered an ignominious breakdown of its huge telephone network when they introduced a new computerised switching system. The software which improved the call switching rate failed at a crucial time and around 40 million calls were not connected. People in the news business found that the problem was itself newsworthy but it could not be communicated to the broadcasters due to the fault.

Here is a case where new software appears not to have been adequately tested. Adequate program testing procedures are a must, particularly when critical functions are dependent upon them. In this case rigorous destructive testing should have been operating.

FEBRUARY 1990

- The publishers of a magazine supporting the Apple Macintosh market were mortified to learn that the 2000 diskettes they had prepared and despatched to their subscribers were infected with the WDEF virus. The diskettes were intended to carry only files containing free font software but unfortunately they carried the virus as well. The publishers had inadvertently imported it into their office on HyperCard software brought from America. They responded quickly and mailed every subscriber a copy of the anti-virus Disinfectant 1.5 program before any serious harm had been done.

  This case indicates just how widespread viruses have become and just how easily it is to import one into a system. Where software developers or magazine publishers offer trial disks of their wares they must ensure their absolute integrity. In this case the publishers took the proper action immediately. In other cases one can envisage victims rushing to law in order to claim substantial damages.

- A young man in Fulton County, Atlanta, Georgia has filed a damages suit against two computer equipment suppliers after terminal equipment blew up in his face causing concussion and multiple scarring. He was working at his terminal when the screen went suddenly blank. He consulted the manual which advised him to switch the computer off for ten seconds and then to switch it back on again. As he followed these instructions there was a whistling noise and the VDU tube exploded violently. Apart from the physical injuries, he is now mentally scarred and cannot work with computers any more.

This is an unusual case as terminal/PC equipment safety is now reckoned to be very good. Without additional information it is difficult to comment further and it must remain a salutary lesson on the vagaries of life in a technological world. It is unfortunate that our source is unable to confirm the name of the equipment manufacturer.

- The Australian Department of Administrative Services avoided catching a nasty virus when they circulated an Invitation to Tender to its potential suppliers on a PC diskette instead of in the usual letter mode. One potential supplier returned the diskette infected with a virus.

  It is not recorded just how this virus came to light but there appear to be no reports of extensive damage or uncontrolled multiplication of the code. When using technology in new ways one must be ready to deal with unexpected accidental results or malevolent manipulation by unscrupulous individuals.

- A new and malicious Trojan Horse type of program has appeared in Canada. It is called 'Mosaic' and runs on Apple Macintoshes. When run it destroys directories of all unlocked disks and diskettes attached to the machine. It then prints 'GOTCHA!' on the screen.

This is a typical malicious program and there are no reports that it has any virus tendencies ie copying itself to other programs. It is fairly simple to identify and delete with commonly used file manipulation software but it still comes as a shock to the unwary.

- A new and positively malignant type of PC virus has been discovered in America. It incorporates sophisticated programming techniques which cause it to remain undetected by the more common virus detection programs. The virus comes in two varieties at the moment, named 1260 and 4096. They are apparently harmless in their present guise and hundreds are thought to be in circulation. American experts, however, are speculating that a third virus will soon be circulated which will transform them into engines of destruction.

  We await further developments in some trepidation. Clearly virus planters are becoming very sophisticated. This approach of circulating the base code in a form which does no damage and then triggering it with another virus is reminiscent of binary biological weapons recently outlawed by the international community. If things get any worse similar action may be necessary with malicious software.

MARCH 1990

- The computer which records trading at the Central Gilts Office in the Bank of England failed on 5th March 1990 for the first time since Big Bang. The fault affected both the main production processor and the back-up computer and led the Bank to issue a warning that some deals were not being properly settled. The fault remained undiagnosed for over 24 hours and users were advised to prepare for paper-and-pencil deal recording until it was rectified.

  Faults which occur simultaneously in main processors and in on-line back-up machines tend to indicate communications problems. It is to the Bank's credit that they have managed to maintain CGO processing continuously since the system was installed. However, it is becoming increasingly apparent that where large-scale processing is required continuously there is likely to be a need to arrange back-up communications as well.

- The new customer support centre operated by the Digital Equipment Corporation (DEC) in Basingstoke has burned down at a cost of £38 million. Although DEC had occupied the site for less than two years, they had properly ensured that disaster relief precautions were in place. A spokesman said that 'There is another support centre down the road which will take on the extra work load. There was very little work lost because everything is backed-up off-site at the end of every day.' However, on reflection he added 'A number of people thought it was a test and are likely to have lost important paperwork.'

DEC have clearly followed the advice they preach and should be congratulated for having up-to-date disaster plans and a 'hot' standby site available to operate the back-up data. Their credibility as a world power in computing terms would have been severely dented if this disaster had caused their customers any serious disruption. However, 'The best laid plans of mice and men gang aft aglay' and those who did not follow instructions when the (genuine) fire alarms went off have unfortunately suffered accordingly.

- A computer printer factory in Spain owned by a major Japanese manufacturer was totally destroyed in just two hours after torrential rainfall caused nearby river banks to collapse. The computer room was flooded to a depth of 10 feet leaving the staff clinging to the roof until rescued. Data recovery at a secondary location took 40 hours.

The vulnerability of this centre to flood damage should have been known to the management. It is not difficult to assess flood-plain boundaries and drainage capabilities in most areas. Either the centre should not have been earmarked for data processing or a more detailed survival plan should have been in place.

- Stormy weather in England resulted in several computer failures:

> - An insurance company in Bournemouth lost all its processing facilities and invoked its back-up procedure - they reverted to pencil and paper routines!
>
> - A software company in Salisbury lost its entire processing facilities when the roof and first floor offices were swept away.
>
> - The uninterruptible power supplies at the Atomic Energy Authority's establishment at Harwell were unable to cope with the load when the mains power failed. Their Cray supercomputer survived but an IBM 3084 and a DEC VAX system lost power.
>
> - An ICL computer operated by a cider maker in Taunton was out of action for half a day although it appears that the company is still unconvinced about the necessity of investing in back-up power supplies.
>
> - A specialist division of a high street bank located in Brighton lost power to its IBM 3081 computer. Apparently the power was restored fairly promptly but staff were unable to locate the switch which controlled the liquid cooling system and processing could not be restarted.

As an incentive to making contingency plans against such disasters, these reports speak for themselves. The recent spate of unusual weather patterns has, apparently, diluted some of the complacency heretofore displayed by UK companies. However, systems managers are still finding it difficult to convince boardroom directors of the need to spend large sums of money on precautions which may never be needed. The recent growth in networking has resulted in the average tolerable breakdown time being lowered from the four days acceptable three years ago to only one day now. Such requirements cannot be achieved satisfactorily unless credible contingency plans are in place.

- In a planned office move, company staff turned up at the new site and found a computer engineer installing a completely incorrect configuration of the corporate computer system.

  This is clearly a case of lack of communication and planning. Sometimes companies seem to view major office moves as little more than week-end activities. In truth they should be considered major projects and be subjected to commensurate planning and implementation controls.

APRIL 1990

- Towards the end of 1989 IBM had identified around 30 different PC viruses. This figure had jumped to almost 80 within the following 3 months.

This information confirms our belief that the incidence of computer virus infestation is likely to rise as time goes on. The complexity and malignity of computer viruses are also likely to increase requiring increasingly stringent controls to be imposed over both physical and logical access to systems.

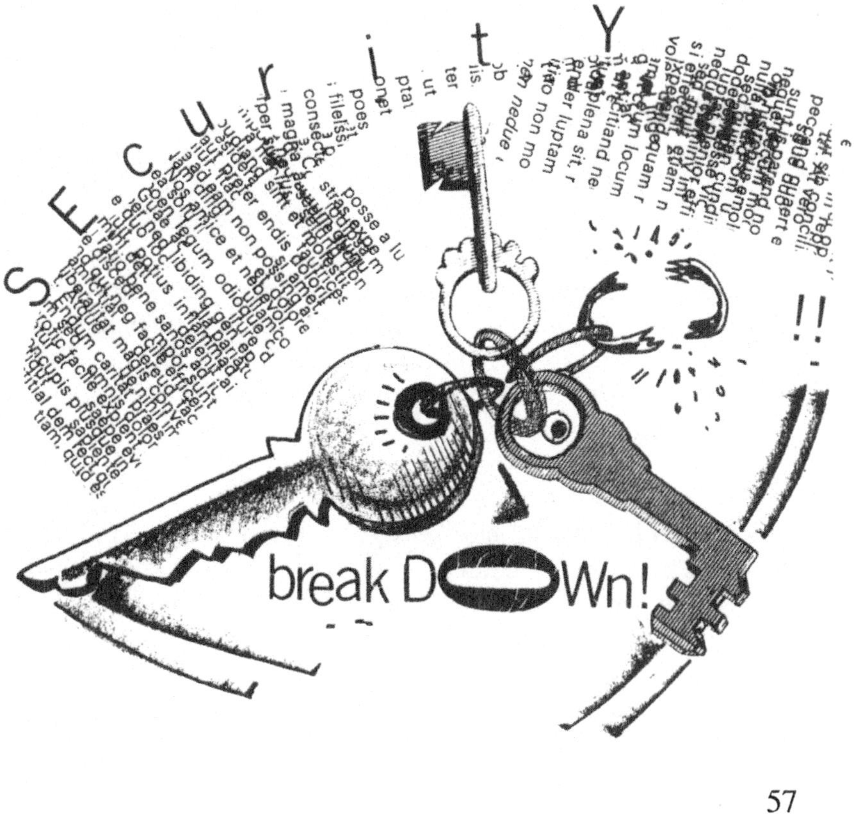

MAY 1990

- The US Secret Service employed 150 agents over 18 months in 15 cities to investigate computer related crimes. Their operation, code named Sun Devil, culminated in the seizure of 42 computer systems, 22,000 diskettes and some guns. Three people were arrested for the firearms offences but nobody for computer crimes. However, prosecutors are still confident that their examination of the diskettes will reveal evidence of criminal activity. They are especially interested in the illegal trading in credit cards and long distance telephone codes. There are also reports that some of those being investigated had illegally hacked into banks and in one instance had entered a hospital computer and threatened life by tampering with patient care data.

  The shape of things to come! Once upon a time Secret Service agents pursued traitors across the world armed only with a Biretta pistol. As computers and the networks connecting them become more complicated and the abilities of the irresponsible become more sophisticated it seems inevitable that greater resources will need to be devoted to preventing such Hi-Tech crimes. It simply illustrates that both organisations and individuals must approach IT security responsibly in order to protect their assets against such attacks.

- A firm of financial advisers got some unwanted publicity when a television detective series was trailed showing the ease of computer crime and featuring a data screen indicating the firm's title. The company took swift legal action and the television company changed the programme to ensure that the target of the 'hacking' was not revealed. It appeared that in dramatising the event they had neglected to check whether their fictitious target actually existed. The company spokesperson complained that showing such untruths were likely to affect the firm's credibility.

Such programmes are appearing frequently now. Usually the scenario includes a pair of, 'bright young brokers' who quickly crack ridiculously simple password systems and help themselves to millions of pounds before being nastily murdered by 'real' criminals. The portrayal of financial computer security is laughably inadequate. However, it appears that it reflects the general public's perceptions. It is, therefore, up to the industry to take steps to ensure that the real situation is not as portrayed and to discourage any attempt by individuals to mimic the media portrayal.

JUNE 1990

- The investigation into a particularly nasty double murder in Wales was brought to a halt because the approved computer system used to manage detective casework did not have the capacity required. The enquiries database grew well beyond the initially anticipated requirements outstripping the record storage capability of the hardware. Transferring the data to upgraded replacement hardware was achieved in record time, under some pressure.

  The 'down' side of this story reveals the problems which can occur when systems analysis seriously underestimates capacity requirements. The 'up' side is that the case management software, bearing the acronym HOLMES (it would wouldn't it?), was designed to run on several different computers and did not need to be replaced.

- Another computer virus, this time called MDEF, which runs on Macintosh Plus desktop computers, has been discovered in a university in New York. All its functions are not yet fully documented but it does cause unpredictable system crashes and data losses.

  The continuing appearance of new viruses in desktop computers has led to the profitable creation of anti-virus software from quick-off-the-mark software houses. Any Macintosh network really should now carry one of these all-embracing virus protection products. Most of the vendors now offer an automatic upgrade service to keep customers protected from new viruses. They often also offer site-licensing facilities so that large numbers of computers can be protected economically.

- The data processing department of a local authority district council in England, which was recently privatised in a management buy-out, collapsed with serious debts and is to be wound-up. Amongst other products, the organisation provided software for fifty district councils for processing the newly imposed Community Charge. They are now seriously worried about software support for the future.

  This event is likely to cast a damper over the government's competitive tendering strategy and the drift of local authorities towards the privatisation of their computer centres. Some advisers suggest that proper Facilities' Management arrangements would improve the council's situations. At least it could get no worse than at present.

JULY 1990

- The national press began ringing alarm bells amongst buyers of a PC magazine when it was reported that the free diskette taped to the front cover of each copy was infected with the 'boot sector' virus program 'Disk Killer'. Thousands of copies of the diskette had been manufactured and circulated. The first to get it were those on the mail order list. However, it was not possible to recall the magazine and the virus also made its way to newsagents magazine stands. It later transpired that the virus was indeed 'Disk Killer' (sometimes known as 'Ogre') but that it was resident in an inactive form.

  The fact that this virus got on to thousands of diskettes being given away free to the public with a magazine is disturbing. It is more than fortunate for the magazine proprietors that the code was inactive for they could have faced serious damages claims had this not been the case. It cannot be long before a live virus will be circulated, either inadvertently or maliciously, over such media. The remedy for IBM-compatible PC boot sector viruses is to use the diskette solely as a code or data source and not to use it as a system diskette. However, this presupposes that its presence can be identified and this can only be accomplished by using a suitable virus detective program that has been kept up-to-date.

- No sooner have the problems over computerisation of Community Charge demands subsided than problems have started arising in software supplied to local authorities for the collection of business rates. The software, which is designed to run on ICL computers appears to cause two serious problems. Firstly the data which is displayed on terminal screens appears to be incompatible with on-line databases and secondly it cannot handle the mailing of bills which has had to revert to manual operation. Several councils have complained to the supplier who has agreed to deal with the situation.

Once again we see problems arising from the introduction of software which appears not to have been tested adequately in a functioning user environment. Time pressures caused by the introduction of national legislation are difficult to combat and users have found themselves having to implement regulations for which there are no adequate tools.

- A large mainframe computer belonging to one of the bigger building societies has been suffering a series of awkward breakdowns. The root cause was the uninterruptible power supply (UPS) causing the mainframe to cease processing. It took the computer vendor's engineers several goes at reloading the system before it stabilised.

  It is all very well having a UPS but if it fails there are likely to be serious consequences for any attached mainframe computers. One of the problems is that they do not like being turned on and off (particularly if this happens rapidly). A UPS is supposed to cover against minuscule breaks in supply. You need a back-up generator with an autostart facility to be covered against all eventualities as UPSs normally can only function from batteries for up to perhaps half an hour, although one did keep going for four hours once.

- After the scares that computer VDU terminals may emit harmful radiation, there are now concerns that the intensive use of a terminal keyboard is far more injurious. One of America's largest telephone companies has just launched a programme costing $8 million which it hopes will reduce injuries amongst staff. They had suffered nearly 700 cases of repetitive strain injury and nearly 300 cases of carpal tunnel syndrome. There are even reports that keyboard manufacturers may face heavy damage claims over product liability.

  The computer is a wonderful and versatile machine. Its productivity has transformed business over the past 20 years. Now we are finding out that people cannot keep up with its demands without widespread injuries. The machine was made to serve man not the other way around. Perhaps it is time to sit back and work out who is supposed to be serving whom!

- The Ministry of Defence has admitted that the computer controlled air defence command and control system ICCS is not operating according to its specification as a fully integrated system. A spokesman in the House of Commons admitted that the situation "...might be said to leave a hole in the defence strategy." The system is so far behind schedule that it is certain to be obsolete before it is commissioned in the mid-1990s.

Large defence systems are complicated to operate, particularly as they function in a real-time environment. The timescale of implementing such systems is so long that the claim of it being obsolete before it is in use is not uncommon. Most large computer systems fall into this category because the speed of development in the industry is so fast.

- A senior London Fire Brigade officer has reported that its computerised command system is unable to cope correctly with the deployment of resources to fire incidents. Although the Brigade states that it is officially satisfied with the system there are reservations about software bugs which are to be fixed immediately rather than left for fixed cycle reviews.

  Apparently the problems were some of the Brigade's own making in that they have changed the specification of the system several times in the last five years. This, of course, also put the price up considerably. There seems to be an uncontrollable urge on the part of many senior managers in many industries to change computer system specifications after they have been published as firm. When this happens several times it is highly likely that the product will fulfil none of them very efficiently.

- It was reported that several employees of British Gas are suspected of carrying out a massive computer fraud. Figures initially banded about indicated losses of up to half a million pounds. Officials at first inspected their computer files and were later reported to be taking days to sort through microfiche records of transactions. Sometime later a much less sensational report indicated that the original suspicions of computer fraud had proved unfounded. Losses had been found amounting to between £5,000 and £10,000 concerning the fraudulent ordering of equipment for private installation work.

This is a classic case of fraud and one of a type which is relatively commonplace in large service organisations. The management retraction of the computer fraud claim was undoubtedly correct if you subscribe to the professional view of the definition of computer crime ie the surreptitious alteration of program stored data. This case almost certainly involved the speed and stupid efficiency of a computer, witness the use of microfiche, but the fraudulent data was probably captured at a legitimate terminal from bogus source documentation. The control of authorisations and source data capture is an anti-fraud requirement which is so often missed when imposing controls over computers in many businesses.

- A computer installation processing financial data for around twenty government departments at the US Department of Agriculture's National Finance Centre has been severely criticised for operating financial controls so poorly that a financial disaster could occur any time. Several audits have been carried out and these indicated that the problems were caused by the enormous number of financial transactions processed. Another reason was the replacement of systems without adequate testing.

This is a case of an apparently successful system being allowed to grow until it became so unwieldy that it was impossible to control properly. The security management of large financial systems is both difficult and expensive and budgetary constraints on the US government are likely to be the root cause of this particular disaster in the making.

- Those involved in the esoteric world of computer-based cryptography were very interested, and possibly slightly alarmed, to read uninformed press reports that the Rivest, Shamir and Adelman (RSA) public key cryptosystem had been cracked. This system has been around for a number of years and is the basis of several commercially successful implementations for the secure electronic transfer of funds between financial computers. The story revolved around two mathematicians who had cracked the Ninth Fermat Number, a factorisation series upon which RSA is based. However the mathematicians had employed massive computers over several months and only succeeded in identifying a single key pair. In computer terms their feat had expended 275 million-instructions-per-second/years.

  The method by which RSA key pairs are generated is a complex process. It is actually quite comforting to discover that even the brightest of brains using the fastest computers take months to crack a single key pair. If appropriate key management procedures are in place the exposure generated by this discovery will be adequately covered. The essential fact proved is that the technology required to generate key pairs is at least a thousand-fold cheaper than that required to crack them.

AUGUST 1990

- The proprietor of a small software business may face the loss of his livelihood after thieves broke into his home in London and stole all his computing hardware and software. They did not take other items of value, just the PC and all the copies of an accountancy package he had been working on for four months. Police believe the culprit was almost certainly a business rival.

  This situation proves the validity of always keeping at least one magnetically recorded copy of valuable software in a safe location some distance from the processing centre and a final paper copy of all documentation at another. Normally one expects such rules to be observed primarily by large commercial and financial organisations. From this episode it is clear that organisations of any size are at risk and need to take proper precautions.

- The spread of Western-type freedoms into Eastern Europe is to be applauded and hailed as the wonder of 1990. However, it appears that some individuals over there do not understand the difference between avoidance of oppression and evasion of responsibilities. An applications software producer in the United Kingdom makes working copies of his products available for free circulation as a marketing ploy. These copies have a routine which causes the screen to display a sales message at four minute intervals. Now a clever computer buff in Poland has found out how to suppress this device and has written to the manufacturers threatening to publish the routine unless he is paid for his efforts. Clearly a sophisticated attempt at business blackmail.

Part of the West's burden must now obviously be introducing our less fortunate neighbours to the differences between enterprise and criminal activity.

An ICL 3980 computer system at the Department of Social Security's Central Operational Strategy centre near Newcastle failed, cutting off all on-line DSS offices from central processing and stranding all new DSS claimants without their benefits. The computer controls an index of every adult citizen in Britain and the problems accessing it caused a domino effect across all the systems which need it. The centre manager blamed a single hardware failure for the problem.

That a single hardware failure can knock out an entire processing network which is essential for the well-being of a significant part of the population is unfortunate to say the least.. It is likely that this case derives from a series of smaller problems which had gone unnoticed by management. There is also the possibility of under-investment in system redundancy due to budgetary constraints.

• A count of the events related in this publication will reveal that we have reported over 100 incidents. These fall into 15 differing types which are summarized below:

TYPE ANALYSIS

Serial	Type	No of Incidents
1.	Poor supervision of staff	1
2.	Line tapping	1
3.	Espionage	1
4.	Communications failure	1
5.	Theft	3
6.	Fault control	4
7.	Other crimes	5
8.	Malicious programming	4
9.	Fraud/embezzlement	6
10.	Line fraud against a financial institution	6
11.	Program testing	12
12.	Hacking	11
13.	Management faults	18
14.	Disaster	16
15.	Virus	21

A discussion of our perceptions of these headings follows overleaf.

Generic Risk by Type

- This section describes the types of generic risks that exist around IT systems into which each of the incidents described in the Type Analysis on the previous page have been categorised.

1. Poor supervision of staff

If staff are poorly supervised there can be little confidence that they are performing as expected. Management need to ensure that staff morale is high and that adequate controls and checks are in place so that unauthorised or malicious staff activity is discovered and can be controlled.

2. Line tapping

As the use of data communication networks grows the risks of sensitive data being read, captured or modified by means of data line tapping increases. Only minimal knowledge and equipment is required to tap a line and the benefits to a fraudster or industrial spy can be immense. The risks of a well placed tap being detected are minute. Protection against line tapping is best performed by the use of cryptography to provide confidentiality and message integrity services.

3. Espionage

While being unpleasant to accept it is a fact that most organisations have spies in their midsts. Staff leave sensitive information on their computer terminals while they go to lunch, printouts are not shredded, dial up access is permitted without adequate safeguards and disks containing confidential information are inadequately protected. Because of these failings espionage either by internal staff working illegally for a competitor or an industrial spy can obtain access to price lists, prospect lists and proposals. Physical access controls, a structured approach to data security and a degree of management awareness to these risks are the best first steps to take in protecting against espionage.

4. Communications failure

Accidental or deliberate failure of communications networks can cause loss of business, unnecessary costs and a potential loss of overall efficiency. For example in an EFTPOS system the failure of the data line between a retailer terminal and an authorisation centre will cause the retailer a loss of trade. These risks can be minimised by adequate network management, back-up data lines and tested manual fallback procedures.

5. Theft

Information is unique among assets in the sense that it can be stolen without being detected. It can be copied or read without detection in most organisations. In addition theft of IT equipment can result in the loss of physical assets and the data stored on it. Protection against information theft is best achieved by means of logical access controls and physically securing the equipment involved. There is a large black market in stolen PCs that have been stolen from organisations. This can be avoided by proper access control systems.

6. Fault Control

This is generally a problem of system manufacturers, whether they are hardware of software based. There have been several instances of manufacturers selling systems which have subsequently required extensive repair due to some unforeseen problems occurring. This can obviously cost the purchaser much time and effort to put right and such expenditure may not always be reimbursed by the system vendor.

7. Other crimes

Crimes involving computers are generally expected to involve either the embezzlement of funds or possibly the destruction of important data. As computers invade more and more areas of activity it is sobering to realise that even our very lives can be at risk. We have, therefore, included in this book a few instances of crime being committed against the person using computers. The commonest of these has been attempted murder via medical computers in hospitals. Our only records of this happening come from the U.S.A. but who is to say that such problems will not occur in Britain?

8. Malicious programming

The risks associated with malicious programming are immense. Disgruntled or criminally inclined programmers can write or modify applications so that they gain financially or the company goes out of business due to decisions made on the basis of information provided by the malicious application. Precautions to decrease the risks of malicious programmers include a proper procedure for application testing, adequate and frequent software audits, staff vetting and configuration management controls.

9. Fraud/embezzlement

This is the immediate activity which springs to mind when considering financial losses associated with computer systems. Often this is trumpeted in the press as 'Computer Crime'. Where the fraudulent data has been captured from a legitimate terminal accessing legitimate software then it should better be described as 'Computer Assisted Crime'. Where the embezzler has the skills to write his/her own programs to access and change data files for personal gain eg reducing the record of a debt or changing the record of ownership, then it is rightly a 'Computer Crime'.

10. Line fraud against a financial institution

Line frauds can be perpetrated against banks because they participate in shared networks used to transmit financial movement instructions. Some systems eg SWIFT have very sophisticated controls to stop misuse. Other, older systems found amongst some European currency exchanges have much weaker systems. If SWIFT is breached then the perpetrator is almost certainly an insider with access to the source documentation which generates the transactions. One of the incidents in this book records an attempt at a large-scale line fraud in a financial institution using simple methods against a European exchange. It was very nearly successful.

11. Program testing

Without rigorous testing of application programmes management can have no confidence that they function as intended. The risks to the organisation caused by a lack of testing include incorrect processing, application unreliability and fraudulent functions being included. In order to decrease these risks management should ensure that application testing is performed against an approved testing specification by staff who did not write the application, emergency "patches" are tested thoroughly and that if appropriate negative testing is performed.

12. Hacking

Hackers gain unauthorised access to IT systems either for fun or for fraudulent gain. Their "efforts" result in information theft, malicious programs being inserted and resource theft. Hacking can be prevented by protecting dial up access, effective password management, proper implementation of logical security systems and encryption of sensitive information.

13. Management faults

The effective management of computer systems is vital for the continued success of any business using IT extensively. This book has been published so that managers may be made aware of the myriad ways in which losses can occur and briefly to draw attention to ways of reducing such risks. The use of IT will spread in business and managers must be in a position to understand the risks and to take effective counter measures in order to protect their enterprises.

14. Disaster

As more and more business critical records and transactions are stored and processed on computers, the long-term viability of the business concerned becomes directly proportional to the investment in disaster recovery plans. A recently published survey of incidents in the U.S.A. tends to indicate that 90% of all businesses which suffered a major computer disaster failed within one year of the incident. Senior management personnel should occasionally ask their data processing managers "If our computer centre caught fire right now, what would be the consequences?" - and then test it in a formal exercise. The results can often be surprising.

15. Virus

There can be few people who have not heard of the computer virus - although significantly fewer can properly describe the beast. We believe that a computer virus must have the capability of either reproducing itself within the host computer or propagate copies of itself across a computer network (ie shows signs of life) before it can properly be described as a virus. Almost all known viruses have occurred on IBM compatible personal computers or on the Apple Macintosh desk-top computer. The proliferation of these machines and their frequent connection to telephone based networks is the reason why they have fallen victim to virus attack so often. Many main-frame computer managers feel that they are immune from such attack. For this reason we have included a fairly detailed account of a computer virus which has been found in DEC VAX equipment. This is surely not the last such incident.

- For a slightly different view, we also offer an analysis of the reported incidents by sector:

SECTOR ANALYSIS

Serial	Sector	No of Incidents
1.	Banking/Finance	21
2.	Aerospace Technology	4
3.	Insurance	4
4.	Hobbyists/Hacking	6
5.	Stockbroking	1
6.	Government/State	15
7.	Law/Police	3
8.	Hardware/Software Manufacturers	8
9.	Commercial general	25
10.	Media	4
11.	Network operators	7
12.	Science	2
13.	Viruses general	15
14.	Railways	2
15.	Accountancy	3
16.	Espionage	1
17.	Shipping	2
18.	Fire brigade	2
19.	Travel agency	1
20.	Medical	8
21.	Criminal extortion	2

THE WIDER VIEW

The incidents which have been reported in this book are all actual events. They serve to illustrate the reality of threats to the assets of organisations and individuals. However, there are many, many more incidents which are not reported because such publicity is regarded by the organisations concerned as unnecessary or even undesirable. Because of this, the overall picture provided by these incidents is distorted from reality.

Therefore, in this section we provided a more complete analysis of the overall situation in respect of risks and threats to assets due to computer systems. In keeping with the theme of this publication - reality and actual events - we have kept this analysis as factual as possible and avoided opinions or subjective views.

The purpose of security measures is generally to protect assets from loss. Ultimately the loss of an asset, disclosure of its nature, or lack of availability will result in financial loss either actual or relative.

Information Assets in Perspective

Assets can be grouped under five headings. They are:

- money

- people

- property

- information

- reputation

Every organisation needs at least two of these assets and very few can manage to operate effectively without all five.

The check-list on the following pages outlines key assets any organisation have that may be compromised by breakdowns in computer security. Readers may find it useful to check the procedures surrounding the security of these assets within their own organisations.

Of the five main asset groups, information is rapidly growing in importance. In many organisations it has already become the most important over the next few years, and there will be a process of adjustment. The first stage in this process is the realisation and recognition of the new status of information.

Properties of Information

One of the most important prerequisites to good security of information is an understanding of its properties. It has several properties in common with other assets:

- it is essential for the effectiveness of the organisation

- its value is very dependent on its quality. Poor quality information is, at best, of no use and can be positively detrimental.

ASSET GROUPS

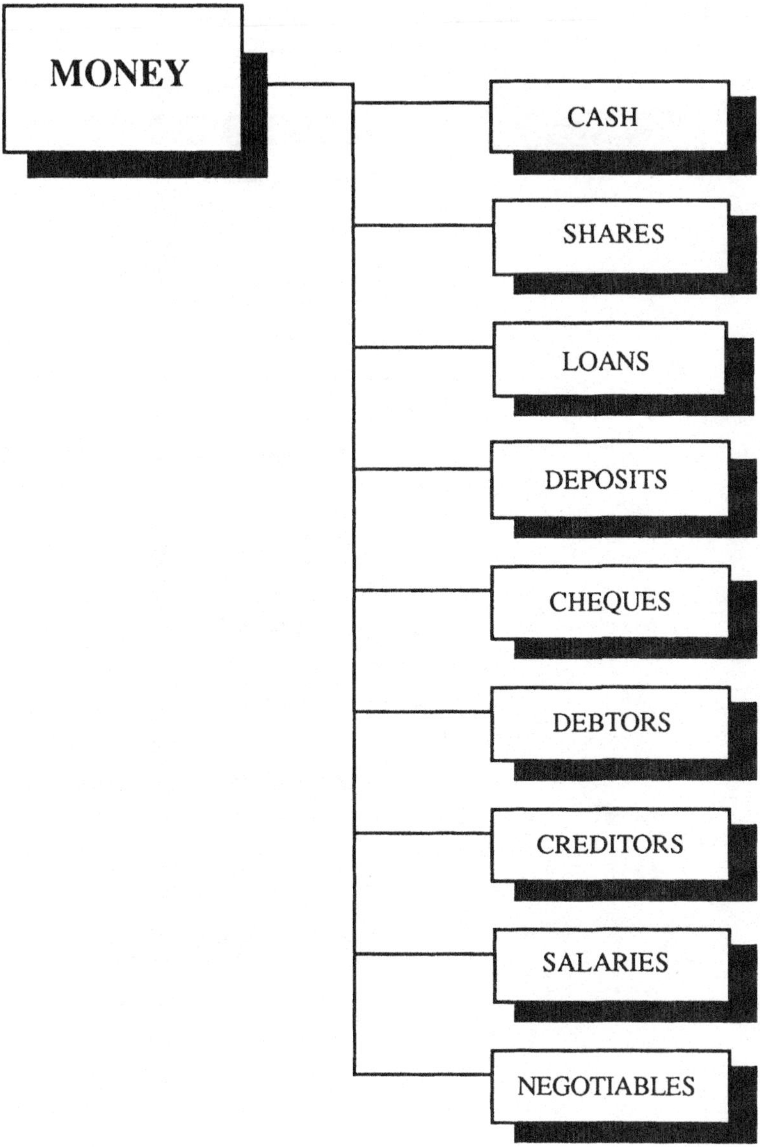

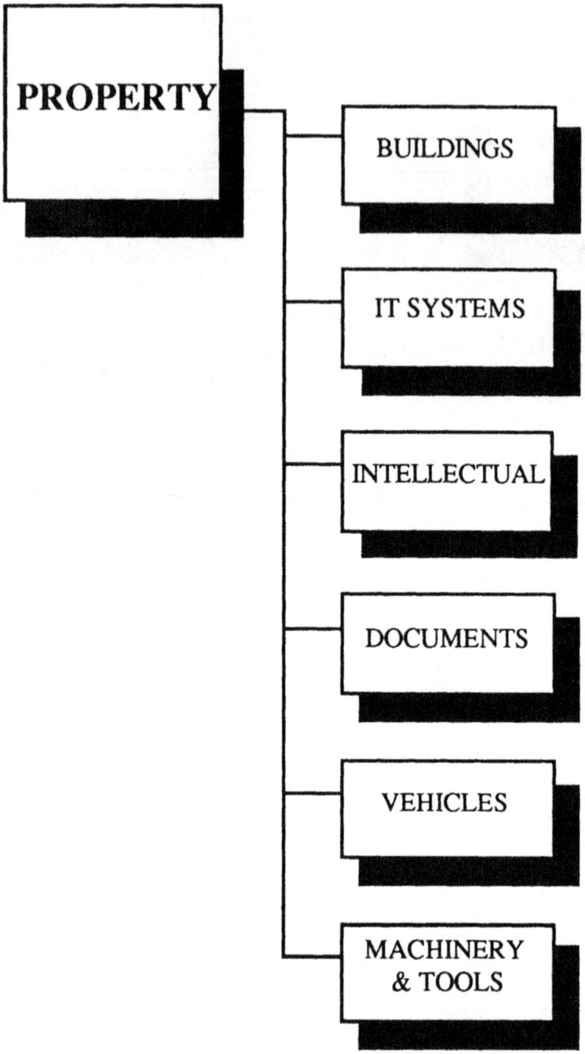
PROPERTY
BUILDINGS
IT SYSTEMS
INTELLECTUAL
DOCUMENTS
VEHICLES
MACHINERY & TOOLS

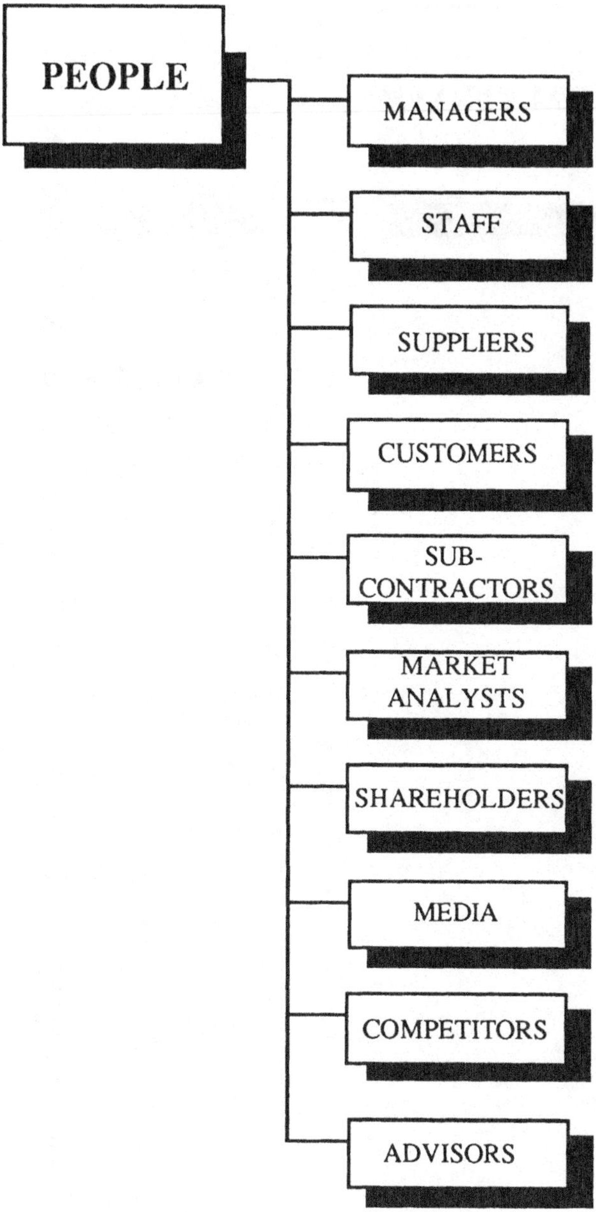
PEOPLE
MANAGERS
STAFF
SUPPLIERS
CUSTOMERS
SUB-
CONTRACTORS
MARKET
ANALYSTS
SHAREHOLDERS
MEDIA
COMPETITORS
ADVISORS

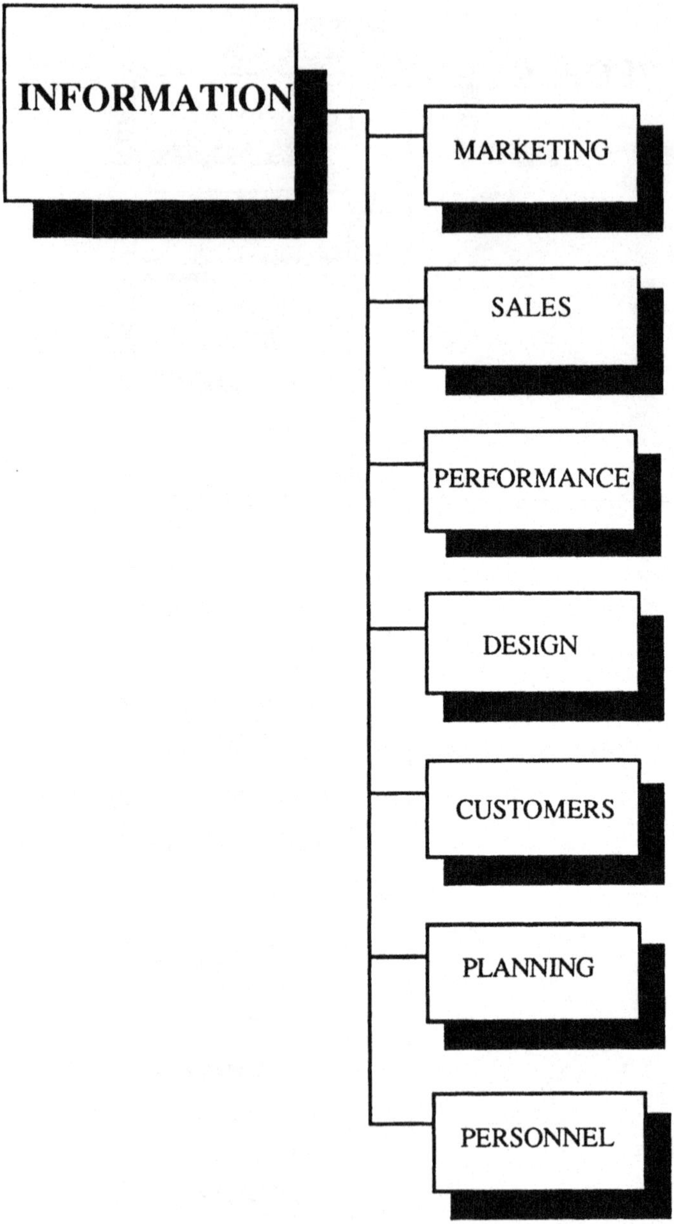
INFORMATION
MARKETING
SALES
PERFORMANCE
DESIGN
CUSTOMERS
PLANNING
PERSONNEL

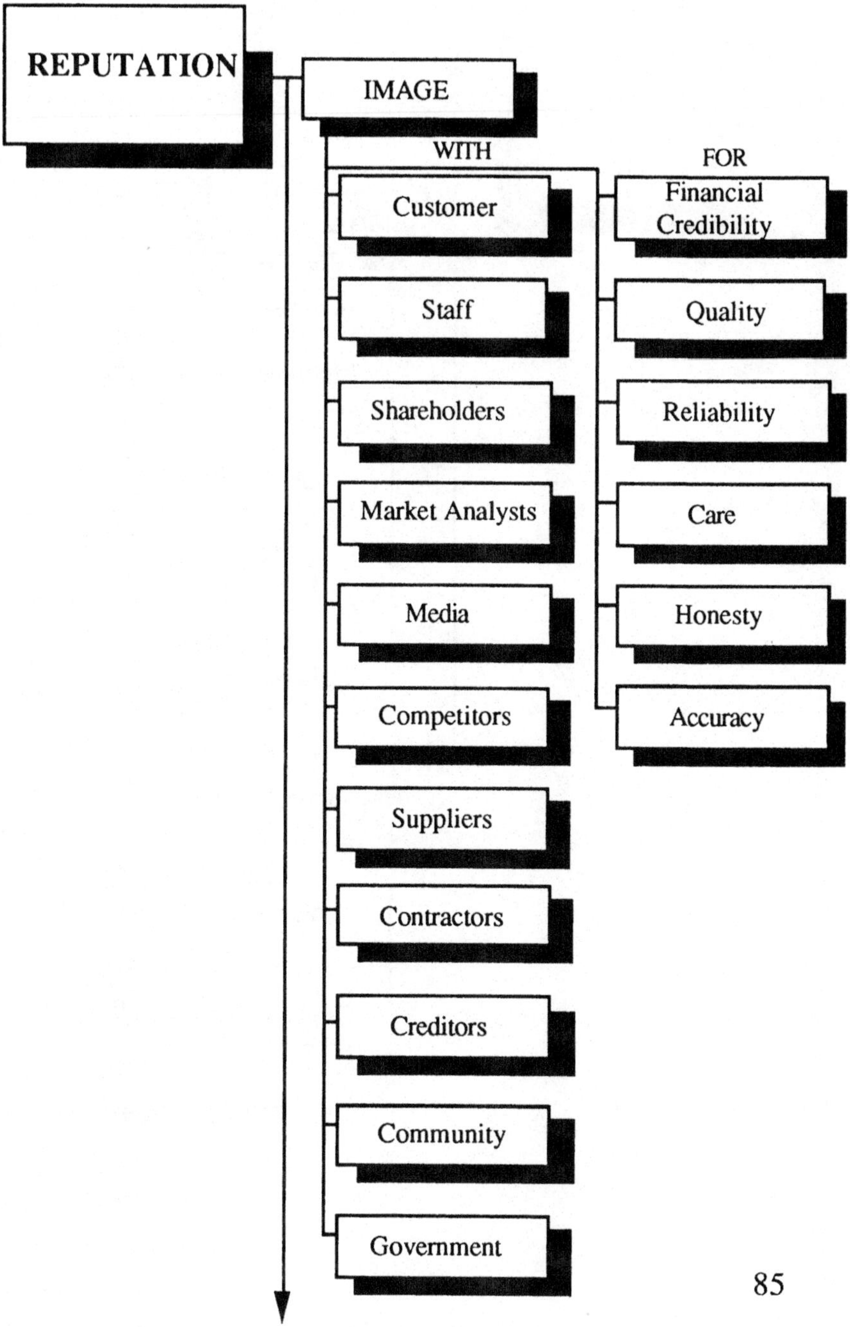
REPUTATION
IMAGE
WITH
FOR
Customer
Staff
Shareholders
Market Analysts
Media
Competitors
Suppliers
Contractors
Creditors
Community
Government
Financial Credibility
Quality
Reliability
Care
Honesty
Accuracy

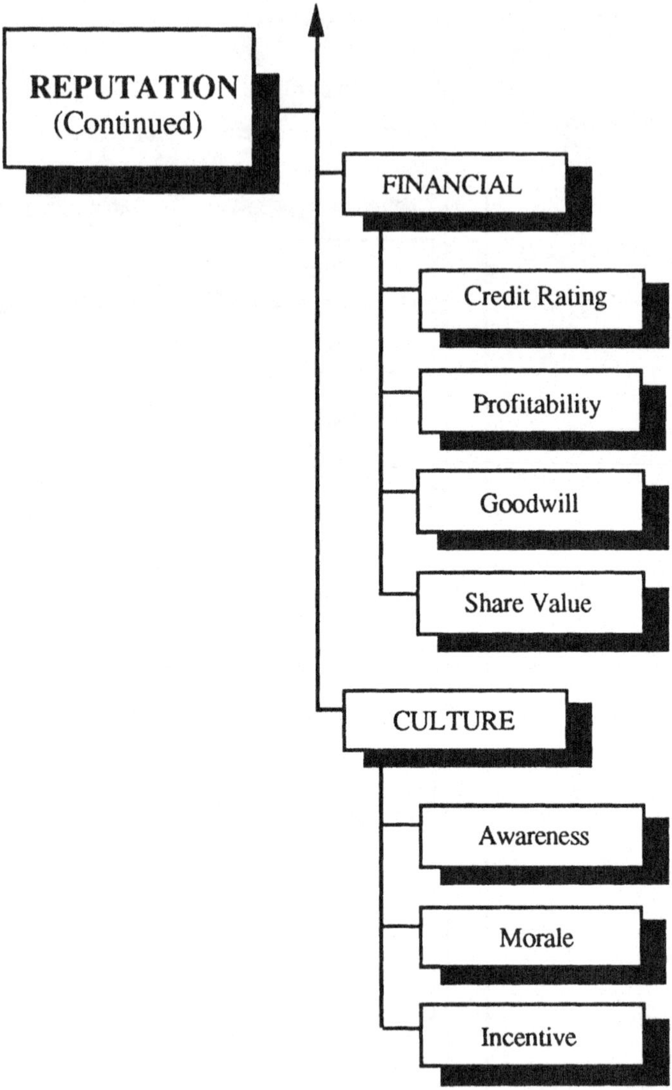
REPUTATION
(Continued)
FINANCIAL
Credit Rating
Profitability
Goodwill
Share Value
CULTURE
Awareness
Morale
Incentive

separate elements of information, when brought together appropriately, are of greater value than the sum of the individual values. This is an important property to remember when defining the classification of information.

Information is also different. It has properties which result in new risks and threats. It also demands a different approach to security.

It is not used up when it is used or stolen. This makes it difficult to detect when it has been stolen and even more difficult to prove it. Furthermore, theft of information is not a crime so one cannot rely on any deterrent value from the law.

Duplication is very easy and cheap. It is almost impossible to prevent duplication by someone who has authorised access to the information.

The value of information is largely independent of quality. A single bit of information can take on any value we choose.

The speed of movement of information is extremely rapid compared to other asset groups and, perhaps, more significant the speed is increasing while the cost is dropping.

Information is almost completely free of government restrictions on movement including transborder flows. This is not the case for people, money or property.

It is extremely simple to change computer based information and to leave no trace at all. Audit trails and the like can give some measure of protection but they are usually in the form of computer based information which can itself be changed. Consequently it is prudent to exercise caution in the extent to which one believes an audit trail.

PA Experience

PA does much consultancy work for clients in the field of IT systems. Security is one of the important aspects and the work has provided a great deal of experience of typical problems which arise. Clearly our work for clients (particularly in this area) is confidential and we will not discuss specific work. However, the wide experience allows us to identify common problems and misunderstandings which will help others to avoid them or at least to be aware of them.

Awareness

The first, and perhaps most important, common finding is a lack of awareness of the subject. The lack exists generally at all levels of seniority but more so at higher levels. It is often the case that the only awareness which is present has arisen through the activities of the media. The result is a distorted perspective based on the sensational (and often exaggerated) issues.

A basic level of awareness of business assets, risks, threats, countermeasures, responsibilities etc at all levels in an organisation is not difficult or costly to achieve. It is the most cost-effective way of improving security for most organisations.

Too Much Paper

There is a marked clinging to old values associated with the use of computer based information. People retain a need to have and keep (unnecessarily) paper copies. Such practices are expensive in the first place but they often also lead to weak links in security or wasted expenditure in computer system security.

Configuration Control

It is not possible to achieve a good level of IT system security unless there is good configuration control. In the light of this it is rather surprising that many organisations have inadequate configuration control. In particular it is often the case that change control is very slack.

Security Breaches

Another common (and surprising) finding is that there are regular security breaches in many organisations. These fall into three categories:

- major events which are well reported internally and which usually have external impact

- events, both major and minor, which are not reported to more senior management. The motive for not reporting is often embarrassment

- events which are not recognised by the organisation as security breaches. These are perhaps the most worrying because it means that the perpetrator can get away with it, and that the problem still exists.

Over-estimation of Cost

It is common for managers to overestimate the cost of good security. This is understandable in many cases because they are presented with all sorts of technical goodies that can be expensive to buy, install and manage. However, a great deal of good security can be achieved by simple things and many of the good practices and procedures result in other savings for companies.

INDEX

'648' virus, 14
'1260' virus, 52
'1701' virus, 14
'1704' virus, 14
'1813' virus, 14, 15, 16, 43
'4096' virus, 52

Accountancy, 78
  computing disasters in, 16
  package, 68
Aegis Mark 7 weapon system computer, 32
Aerospace technology, 78
AIDS databases, 29
'AIDS Information' package, 47
Alaska, 19
Anti-virus software, 34, 50, 60
APACS computer-based payments system, 44
Apple Macintosh, 50, 51
Assets, awareness of, 89
AT & T, 48
Atlantis space shuttle, 20
Atomic Energy Authority, 55
Audit trails, 88
Australia, 51
Australian Stock Exchange, 33
Authorisations, control of, 66
Auto-pilot system, computerised, 19
Automated selling, 19
Awareness, of security issues, 88

Backup
  communications, need for, 53
  copies, destroyed by systems failure, 23
  of critical data, 30
  system, 23
Bank of Crete, 26
Bank of England, 53
Banking, 78
  fraud, 26
  secrecy regulations, 29
Banks, 2, 17, 33, 43, 44, 48, 74
Belgium, 5, 12
Blackmail, 29, 30, 69
'Brain' virus, 14
British Airways, 35
British Gas, 65
British Rail, 15
Building societies, 26, 43, 63
Business rival, 68

California Student Loan Finance corporation, 15
Canada, 51
Capacity
  management enquiries, 15
  planning, 19
  requirements, underestimated, 60
Carpal tunnel syndrome, 64
Carrier networks, 10
Cellular telephone, fraud, 45
Central Gilts Office, 53
Chaos Computer Club (Germany), 14
  China, 43
  Chips, computer, 9
  City of London, 22, 24
  Clerical Medical insurance company, 24
CMOS-based chips, 31
Collusion, 26
Commercial cases, 78
Communications
  failure, 70, 72
  links, 19
Community Charge, computerisation of, 61, 62
Companies Registration Office (Cardiff), 19
Competitor, attack by, 29
Component testing, 30
Computer 'hygiene', 17, 41
Computer
  crime, 43, 74

TV series showing ease of, 58
type of, 70, 72
disaster, 76
failure, 15
game, 37
related crimes, 58
systems, unstable, 33
Computer Weekly magazine, 27
Configuration control, 3, 89
Contingency plans, 15, 23, 26, 38, 56
Control commands, file of, 42
Copyright legislation, 36
Cost of security, overestimation of, 90
Credit Suisse, 9
Criminal extortion, 78
Cryptography, 67, 71

Data
communications networks, 71
files, corrupted, 7
Database
destroyed by systems failure, 23
outgrew storage capability, 60
on AIDS, accessed by hackers, 29
'Datacrime' virus, 33, 41
DCL commands, 42
Death, caused by hackers, 30
DEC, *see* Digital Equipment Corporation
Defence systems, 65
Department of Agriculture's National Finance Centre (USA), 66
Department of Social Security (DSS), 69
Detection, of unauthorised access to computer systems, 27
Development costs, underestimating, 34
Digital Equipment Corporation (DEC), 41, 53, 77
Disaster
computer, 70, 76
planning, 25, 28
recovery contract, 10
recovery plan, 2, 76
recovery procedures test, 26
recovery site, 25
relief precautions, 53
scenarios, need to take wide view, 39
simulation of, 28
Disinfectant 1.5 program, 50
Disk drives, 30
'Disk killer' virus, 62
Diskette, 47, 62
Docklands Light Railway, 39
Drugs database, tampered with, 31
DSS, *see* Department of Social Security
Dual authorisation, 5

Earthquake, 20, 28, 46
Eastern Europe, 68-69
EFTPOS system, 72
Eire, 38
Electricity supply, failed, 26, 39
Electronic mail
box, 5-6
systems, 7
electronic transfer of funds, 67
Embezzlement, 70, 74
Employees, disgruntled, 4, 39
Espionage, 70, 71-72, 78
EUCLID, *see* Eurobond information and clearing service, 4
Eurobond computer fraud, 4
Eurobond information and clearing service (EUCLID), 4
Export documentation, 22
Extel Financial, 22
Extortion, 29, 47, 78
Exxon Valdez oil disaster, 19, 32

Facilities' Management arrangements, 61
Fault control, 70, 73
Federal Home Loan Bank, 2
Finance, 78
Financial
data, processing of, 66

institution, line fraud against, 70, 74
loss, 79
transaction systems, 6
Fire, 10, 53, 65
brigade, 78
control system, 44
Flood damage, 22, 54
France, 5, 29, 34, 39
Fraud, 4, 6, 9, 17, 26, 34, 37, 45, 48, 65, 66, 70, 74
'Friday 13th' virus, 14, 15, 33, 41, 43
Futures Exchange (New Zealand), 11

Games, computer, 21, 37
General Accounting office (USA), 35
Generic risk by type, 70-78
Germany, 14, 18
hackers in, 35
Government, 78
Greece, 26

Hackers, 7, 13, 18, 29, 30, 37
amnesty offered to, 24
Hacking, 4, 12, 27, 31, 59, 70, 75, 78
laws against, 11, 12, 14, 36
Hardware, 60
buying right, 34
failure, 69
manufacturers, 78
Harwell, 55
Hepnet, 41
Hill Samuel, 6
Hillsborough Stadium disaster, 25
Hobbyists, 78
Holland, 41
'Holmes' software, 60
Hospital
attempted murder by hacking, 31
medical records, accessed by hackers, 30
Host emulation program, 7
HyperCard software, 50

IBM, 7, 12, 30, 57
ICCS control system, 64
ICL 3980 computer system, 69
Information
as asset, 79, 80, 84
properties of, 80, 87
Injuries, allegedly from VDU terminals, 35
'Insider', security threat from, 17, 33, 40, 74
Insurance, 78
Internet, 8, 36
'Italian' virus, 14

Japan, 8

Keyboard injuries, 64
KGB, tricked by German hackers, 18

Law, 78
Laws, against hacking, 11, 12, 14, 36
'Lazy Harry' virus, 21
Lightning, damage caused by, 31, 38
Line fraud, 70, 74
Line tapping, 70, 71
Livermore National Laboratory (USA), 12
Local authority, data-processing, 61
Logic bombs, 4, 8
Logical access control, 5
London Fire Brigade, 65

Macintosh Plus, 60
Macro-Marketing, 9
Magazine proprietors, and virus, 62
Mainframe virus, 42, 77
Malicious
programming, 70, 73
software, 52
Management
awareness, 16
faults, 70, 75
problem, 11, 17, 32, 36

'MDEF' virus, 60
Media, 78
Medical
cases, 78
control systems, 31
databases, hackers gained access to, 29
field, computer-related problems in, 30
records, accessed by hackers, 30
Meteorological Office, 34
Mexico, 31
earthquake, 20
Microchips, 9
Ministry of Defence, 64
Mitsubishi International Finance Corporation, 4
Money, as asset, 79, 81
Morris, Robert Tappan Junior, 8
Mortgage borrowers, service to, 43
'Mosaic' trojan horse, 51
Murder, attempted by hacking, 31, 73

National Aeronautics and Space Administration (NASA), 7, 41
National Security Agency (USA), 8
NATO security, 35
Network operators, 78
New Zealand, 11

Off-site, critical data to be kept, 23
Office moves, 56
'Ogre' virus, 62
On-line search facilities, 19

'Pakistan' virus, 8
Paper copies, 89
Passwords, 5, 6, 7, 9, 31, 42
People, as asset, 79, 83
Personal computers, 7
'Pirated' software, 17, 37, 41
Planning, insufficient, 11
Poland, 68
Portugal, 41
Power supply, 38
Pre-test check, 26
Prime computer networks, 7
Program implementation controls, 4
Program testing, 32, 35, 49, 70, 75
Programming
inflexibility, 21
malicious, 70, 73
proper, 44
Property, as asset, 79, 82
Public systems, 30

Railways, 78
Real-time
environment, 65
medical control systems, 31
Record storage capability, 60
Recovery plan, 10
Remote computing, 20
Repetitive strain injury, 64
Reputation of company, as asset, 80, 85-86
Reuters, 11
Risks, awareness of, 89
Rivest, Shamir and Adelman (RSA) public key cryptosystem, 67
Royal National Institute for the Blind, 43
RSA, *see* Rivest, Shamir and Adelman

Satellite links, 20
Science, 78
Scotland, 11
Seattle, 2
Secret Service (USA), 58
Security (negative) testing, 3
Security breaches, 89-90
Sendmail facility, 8
Senior executives, personally liable for computer disasters, 28
Shearson Lehman Hutton, 4
Shipping, 78
Singh, Edward Austin, 7
Smoke damage, 24
Software

anti-virus, 21, 34, 60
copies kept off-site, 68
defective, 35
licence fees, 47
malicious, 52
manufacturers, 78
not tested adequately, 63
'quarantine', 17
support, 61
South Africa, 44
Soviet
cosmonauts, 3
space exploration disaster, 2
Soviet Union, 18
Space exploration disaster, 2
Spain, 24, 29, 54
Staff
disgruntled, 4, 39
immature, 40
supervision, poor, 70, 71
Span network (NASA), 41
State Earnings Related Pension Scheme (SERPS), 21
Stockbroking, 78
'Stoned' virus, 14
Storm damage, 31, 54, 55
'Super user' status, 13
Sweden, 33
SWIFT network, 74
Switzerland, 41
System interlocks, 3
System specifications, changed, 65
Systems failure, 23

Tapes, accidentally deleted, 32
Tax evation, 29
Telephone
cellular, and fraud, 45
network, breakdown, 48
switching station, destroyed, 10
Television detective series, and hacking, 58-59
Testing problem, 37
Theft, 70, 72
'Trap-doors', 8, 36, 42
Travel agency, 78
destroyed by hackers, 29
Trial disks, 50
Trojan Horse, 47, 51

UDP, *see* Uninterruptible power supply
Uninterruptible power supply (UPS), 63
Union Bank of Switzerland, 9
United States of America, 2, 7, 8, 10, 12, 35, 37, 41, 45, 46, 48, 50, 52, 58, 60, 66, 67, 73, 76
University of Southern California, 12
USSR, hackers and, 35

'Vaccine' program, against viruses, 16, 24, 41
'Value' of information, 18
Vandalism, 15
VAX computers, 41, 77
VDU, *see* Visual display unit
Virus, 8, 11, 12, 15, 21, 24, 41, 42, 50, 51, 52, 57, 62, 70, 76, 78
detective program, 62
free, ensuring software is, 17
infection, preventing, 41
mainframe, 42, 77
planters, 52
triggering one with another, 52
'vaccine' program, 16
writing kits, 14
Visual display unit (VDU)
terminals, injuries allegedly caused by, 35
tube, exploded, 51

Wales, 60
WDEF virus, 50
Weapon system computer, 32
Weather forecasting, 34
Wiring safety, 31

COMPUTER WEEKLY PUBLICATIONS

Computer Weekly is the UK's leading weekly computer newspaper which goes to over 112,000 computer professionals each week. Founded in 1967, the paper covers news, reviews and features for the computer industry. In addition, *Computer Weekly* also publishes books relevant to and of interest to its readership.

Publications to date (obtainable through your bookshop or by ringing 081-685-9435/081-661-3050) are:

The Computer Weekly Guide to Resources 1990

Our extensively indexed second Annual Guide fulfils the computer industry's need for an independent, handy, up-to-date reference review signposting and interpreting the key trends in the computer industry.

A key section this year is an in-depth independent discussion of 270 software and computer companies, invaluable for managing directors, DP managers, sales and marketing people and all executive job hunters.

Our first Annual Guide was well acclaimed: 'In spite of a plethora of guides to various aspects of the computer industry, there hasn't been one readable, comprehensive overview of the current UK scene. *Computer Weekly*'s Guide to Resources has filled the bill ... it's very good.' *The Guardian*

ISBN 1-85384-017-3 416 pages A4 size Price £45.00

Aliens' Guide to the Computer Industry
by John Kavanagh

In a lucid and light style, leading computer industry writer John Kavanagh discusses how the various parts of the computer industry inter-relate and what makes it tick. Complete with extensive index, the book is invaluable for all who come into contact with the computer industry.

'Business professionals who worry about their grasp of the general computing scene and do not want to be bombarded with jargon and technicalities, will get good value ... an excellent 'snapshot' of the companies, the current areas of interest and the problems' *Financial Times*

ISBN 1-85384-012-2 192 pages A5 size Price £9.95

Computer Jargon Explained
by Nicholas Enticknap

Following reader demand this is a totally revised, expanded and updated version of our highly successful guide to computer jargon, *Breaking the Jargon*.

This 176 page book provides the context to and discusses 68 of the most commonly used computer jargon terms. Extensively cross-indexed this book is essential reading for all computer professionals, and will be useful to many business people too.

'... a useful shield against the constant barrage of impossible language the computer business throws out' *The Independent*

'... a worthwhile investment' *Motor Transport*

ISBN 0-85384-015-7   176 pages   A5 size   Price £9.95

What To Do When a Micro Lands On Your Desk
by Glyn Moody and Manek Dubash

This book will help you get the most out of your microcomputer. It is a practical book, giving advice on how to make the transition from typewriter to micro profitably and with minimum effort.

The authors look at software - wordprocessing, databases, spreadsheets, graphics and communications - and the different types of hardware on the market. The book contains valuable information on training, health and security, and legal matters including the Data Protection Act, operating systems, the history of the computer, the current micro scene and the future.

ISBN 1-85384-011-4   296 pages   A5 size   Price £14.95

Considering Computer Contracting? by Michael Powell

This is a completely revised and updated edition of the highly successful book which has helped many computer professionals break loose from being employees to working freelance, in some cases doubling their salaries.

There is information on: who uses computer contractors and why; what it takes to become a contract worker; how to find your first contract; how to keep your skills updated; forming your own company and handling finances; contract agencies.

ISBN 1-85384-022-X   176 pages   A5 size   Price £12.95

Hitchhikers' Guide to Electronics in the '90s by David Manners

Developments in electronics underpin not only the computer industry but also the whole of modern society. This book is essential if DP and IT professionals are to identify trends that will affect all our jobs in the 1990s.

David Manners, an awarding winning senior editor on *Electronics Weekly*newspaper, lucidly explains the electronics industry and its key products and discusses its central role and implications to industry in the 1990s.

Essential reading for IT staff, marketing and sales directors, strategic planners and all interested in the future of the IT industry.

ISBN 1-85384-020-3   224 pages   A5 size   Price £12.95

A Simple Introduction to Data and Activity Analysis
by Rosemary Rock-Evans

Successful analysis of business operations is a prerequisite to building any computer system within a company. Whereas many existing books approach this topic from an academic point of view, this one is the fruit of years of practical analysis in blue chip companies.

Rosemary Rock-Evans is a leading consultant. Her first book on this topic for *Computer Weekly*, published in 1981, is now out of print. However, the considerable demand within the industry for this book has resulted in this totally revised and updated version.

This book is essential reading for all analysts in the computer industry, and is also recommended for students to give them a taste of the real world of analysis.

ISBN 1-85384-001-7   272 pages   A4 size   Price £24.95

Open Systems: The Basic Guide to OSI and its Implementation
by Peter Judge

We recognise the need for a concise, clear guide to the complex area of computer standards, untrammelled by jargon and with appropriate and comprehensible analogies to simplify this difficult topic. This book, a unique collaboration between *Computer Weekly* and the magazine *Systems International*, steers an independent and neutral path through this contentious area and is essential for users and suppliers and is required reading for all who come into contact with the computer industry.

ISBN 1-85384-009-2   192 pages   A5 size   Price £12.95

IT Perspectives Conference: The Future of the IT Industry

Many nuggets of strategic thought are contained in this carefully edited transcript of the actual words spoken by leading IT industry decision makers at *Computer Weekly*'s landmark conference held late in 1987. The conference was dedicated to discussing current and future directions the industry is taking from four perspectives: supplier perspectives; communications perspectives; user perspectives and future perspectives.

'... makes compelling reading for those involved in the business computer industry' *The Guardian*

'... thought-provoking points and some nice questions put to speakers at the end' *Daily Telegraph*

ISBN 1-85384-008-4 224 pages A4 size Price £45.00

Computer Weekly Book of Puzzlers
Compiled by Jim Howson

Test your powers of lateral thinking with this compendium of 187 of the best puzzles published over the years in *Computer Weekly*. The detailed explanations of how solutions are reached make this a useful guide to recreational mathematics. No computer is needed to solve these fascinating puzzles.

'... a pleasant collection of puzzles exercises for computer freaks. Actually probably fewer than half the puzzles here need a computer solution ...' *Laboratory Equipment Digest*

ISBN 1-85384-002-5 160 pages A5 size Price £6.95

Women in Computing by Judith Morris

Written by a respected former editor of several computer magazines, this book reflects the upsurge in awareness of the important role women can play in helping to stem the critical skills shortage within the computer industry.

The book addresses women's issues in a practical and sensible way and is aimed at all business women both in the computer industry or who work with computers. Contains much practical advice, including the names and addresses of useful organisations, and a valuable further reading list.

ISBN 1-85384-004-1 128 pages A5 size Price £9.95

How to Get Jobs in Microcomputing
by John F Charles

As micros proliferate, opportunities for getting jobs in this area are expanding rapidly. The author, who has worked with micros in major organisations, discusses how to get started in microcomputing, describes the different types of job available, and offers tips and hints based on practical experience.

ISBN 1-85384-010-6   160 pages   A5 size   Price £6.95

Low Cost PC Networking
by Mike James

The whole area of PC networking is taking off rapidly now. Can you afford to be left behind? Mike James' book shows how networking revolutionises the way we use PCs and the tasks that they perform. It also explains how networking goes further than simply linking PCs, and how it enables you to integrate your operations to transform your business. Chapters cover every aspect of networking, from planning your network and selecting the hardware and software to applications, technicalities and contacts.

ISBN 0-434-90897-5   256 pages   246 x 188 mm   Price £16.95

Selling Information Technology: A Practical Career Guide
by Eric Johnson

Selling in IT requires more skill and creativity than selling in any other profession. This handbook explains why and provides practical down-to-earth advice on achieving the necessary extra skills. A collaboration between *Computer Weekly* and the *National Computing Centre*, this book discusses career issues, general IT sales issues, and key IT industry developments.

ISBN 0-85012-684-3   244 pages   144 x 207 mm   Price £12.50

Managing Information Security: A Non-technical Management Guide
by Ken Wong and Steve Watt

This book has been written by experienced consultants in what is for most people a new field. Management issues are covered in detail. Topics include: people - are they assets or liabilities, risk assessment, devising and testing a disaster recovery plan, encryption and communication security, the impact of the PC revolution, access control, combatting hacking and viruses, and security in different vertical market sectors such as banking and retail.

ISBN 0-946395-63-2   336 pages   277 x 214 mm   Price £85

Printed in Dunstable, United Kingdom

91222507R00060